Thin Films Photovoltaics

Edited by Beddiaf Zaidi and Chander Shekhar

Published in London, United Kingdom

IntechOpen

Supporting open minds since 2005

Thin Films Photovoltaics
http://dx.doi.org/10.5772/intechopen.96073
Edited by Beddiaf Zaidi and Chander Shekhar

Contributors
Akin Olaleru, Olasoji Adekoya, Joseph Kirui, Eric Nnditshedzeni Maluta, Kenza Kamli, Zakaria Hadef, Muhammad Aamir Iqbal, Maria Malik, Wajeehah Shahid, Juan Felipe Montoya, Esteban Velilla Hernández, Juan Bernardo Cano Quintero, Franklin Jaramillo Isaza, Iván Mora-Seró, Kaiwen Sun, Fangyang Liu, Xiaojing Hao, Beddiaf Zaidi, Chander Shekhar, Syed Zaheer Ud Din, Nadia Anwar, Mujtaba Ikram, Faryal Idrees

First published in London, United Kingdom, 2022 by IntechOpen
IntechOpen is the global imprint of INTECHOPEN LIMITED, registered in England and Wales, registration number: 11086078, 5 Princes Gate Court, London, SW7 2QJ, United Kingdom
Printed in Croatia

British Library Cataloguing-in-Publication Data
A catalogue record for this book is available from the British Library

Additional hard and PDF copies can be obtained from orders@intechopen.com

Thin Films Photovoltaics
Edited by Beddiaf Zaidi and Chander Shekhar
p. cm.
Print ISBN 978-1-83969-905-4
Online ISBN 978-1-83969-906-1
eBook (PDF) ISBN 978-1-83969-907-8

For EU product safety concerns:
IN TECH d.o.o., Prolaz Marije Krucifikse Kozulić 3, 51000 Rijeka, Croatia, info@intechopen.com or visit our website at intechopen.com.

Meet the editors

Dr. Zaidi is an associate professor in the Department of Physics, University of Batna 1, Algeria. He obtained a Ph.D. in Physics at the University of Annaba, Algeria, in 2014. He has published several research papers in reputed journals and written three books. He was editor in chief of the *International Journal of Materials Science and Applications* (IJMSA) as well as an editorial board member for numerous other journals and lead guest editor of many special issues. He is also a journal peer reviewer. Dr. Zaidi has participated in many international conferences.

Dr. C. Shekhar is an assistant professor in the Department of Physics, Amity School of Applied Sciences, Amity University Haryana, India. He obtained a Ph.D. from the Department of Materials Engineering, Indian Institute of Science, Bangalore, in 2012 and thereafter worked as a postdoctoral fellow at the Department of Mechanical and Industrial Engineering, Concordia University, Montreal, Canada. He obtained his MSc in Physics and MTech in Materials Science from Barkatullah University, Bhopal, India. Dr. Shekhar has published thirty-seven research papers in national and international journals and conference proceedings, filed one Indian patent, and presented papers at numerous conferences. His research interests include phase equilibria and thermodynamic properties of ceramics and alloys; CALPHAD modelling, nanostructured thin films for gas sensors, photovoltaics, and dielectric properties of materials.

Contents

Preface

Silicon-based monocrystalline and multicrystalline solar cells are by far the most widely used solar cells today. Thin film technologies in general, and cadmium sulfide (CdS) and cadmium telluride (CdTe) in particular, are increasingly being employed. Other solar cells based on dyes or organic materials are still in their infancy but promise a bright future for more efficient photovoltaic technologies.

In photovoltaics, "thinness" generally refers to so-called first-generation, high-efficiency silicon cells, which are manufactured from bulk wafers hundreds of micrometers thick. Thin films sacrifice some light-gathering efficiency but use less material. In copper indium gallium diselenide (CIGS) solar cells, the efficiency tradeoff is less severe than in silicon. The record efficiencies for thin film CIGS cells are slightly less than those of CIGS for lab-scale top performance cells. In 2008, CIGS efficiency was by far the highest compared with those achieved by other thin film technologies such as CdTe photovoltaics or amorphous silicon (a-Si). CIS and CGS solar cells offer total area efficiencies of 15.0% and 9.5%, respectively. In 2015, the gap with the other thin film technologies was closed, with record cell efficiencies in laboratories of 21.5% for CdTe (FirstSolar) and 21.7% for CIGS (ZSW).

One material with a larger bandgap than Si is perovskite, which has been researched since 2009 for use in thin film solar cells. Starting with a 4% cell efficiency, perovskites have shown rapid progress, demonstrating a 10.9% efficient cell mid-2012 and increasing to as much as 17.9% in 2014. Perovskites are materials following the formula ABX3 where A and B are cations of different sizes and X is an anion.

Thin film-based solar cells that have the common processes for deposition of semiconducting materials with low thickness on different substrates giving uniform appearance produce modules of slightly lower efficiency. The market share for all these technologies is around 10% and remains relatively stable.

Beddiaf Zaidi
Faculty of Matter Sciences,
Department of Physics,
University of Batna 1,
Batna, Algeria

Chander Shekhar
Department of Physics,
Amity School of Applied Sciences,
Amity University Gurgaon,
Haryana, India

Chapter 1

Introductory Chapter: Thin Films Photovoltaics

Beddiaf Zaidi and Chander Shekhar

1. Introduction

1.1 Overview on thin films photovoltaic technologies

Currently, the ever-increasing energy demand has put a strain on conventional energy sources resulting in a catastrophic increase in greenhouse gases and their effects a great cause of concern. This rapid growth in the energy-intensive hymen activities and the ill effects of use of conventional energy resources has pushed scientists to explore and use new renewable, non-polluting energies sources to meet the energy demand of society. Among the renewable energy resources are solar energy produced by photovoltaic panels, wind, geothermal, biomass, hydropower, etc. [1–6].

To tap the abundantly available solar energy especially in the tropics, among the variety of the materials employed amorphous thin-film silicon (a-Si) based solar cells has been mainly used [7]. Among other materials, CdTe (heterojunction cadmium telluride, cadmium sulphide), CIS (heterojunction of copper diselenide indium/cadmium sulfide) have also been used [8, 9]. The manufacturing of these materials is carried out largely in automated systems free or with minimal hymen interventions which is suitable for large-scale productions (**Figure 1**).

In addition to the traditional technologies to generate energy from renewable energy sources, many other approaches and avenues are currently opening up, with a lot of uncertainty that which of these may prove to be more effective over the other [10].

At present, crystalline technologies (multicrystalline and monocrystalline) are by far the most widely used. But "thin film" technologies in general and CIS and CdTe in particular, are increasingly developing being employed by more and more people. Whereas other approaches employing dyes or organic materials are still in their infancy. These approaches though promise a bright future for solar energy generation by photovoltaic cells. There are currently three large families/generations of solar cells as follows [11–14].

- Crystalline technologies which use flat cells from 150 to 200 μm, i.e., 0.15 to 0.2 mm thickness, cut from an ingot or a brick obtained by fusion and molding, then connected in series to each other to be finally placed and glued to the rear face of the module's protective glass. The raw material is always silicon to result in monocrystalline silicon, multicrystalline silicon or amorphous silicon.

- For more details on the manufacturing process of the solar cells, thin film solar cells are based on the use of extremely thin layers of a few microns thick and deposited under vacuum on a substrate (glass, metal, plastic, etc.) A thin uniform layer composed of single or more often several materials deposited through vapor phase. The most developed industrially useful materials are CdTe and CIS-based solar cell technologies.

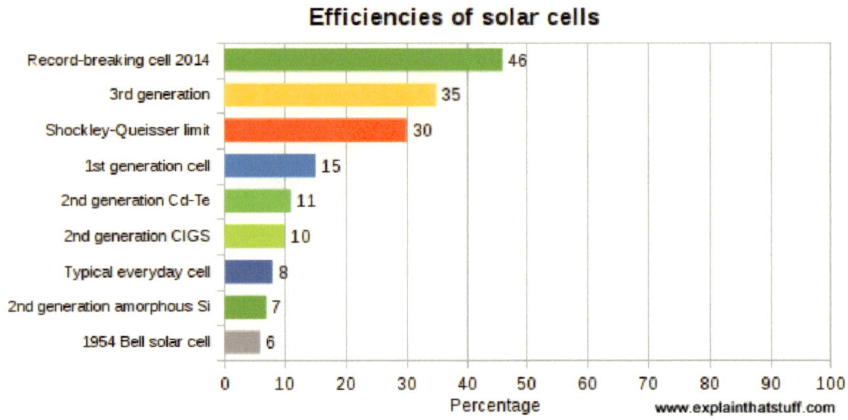

Figure 1.
Efficiencies of solar cells.

- Photovoltaic polymer cells will be mentioned for information as their performance can still only be observed under experimental conditions and they only constitute a long-term alternative. We will also consider concentrators intended for power stations, intended to follow the solar race and permanently expose extremely expensive cells and with very high efficiency.

Author details

Beddiaf Zaidi[1*] and Chander Shekhar[2]

1 Faculty of Matter Sciences, Department of Physics, University of Batna, Batna, Algeria

2 Department of Physics, Amity School of Applied Sciences, Amity University Gurgaon, Gurugram Haryana, India

*Address all correspondence to: zbeddiaf@gmail.com

References

[1] Salameh ZM, Dagher F, Lynch WA. Step-down maximum power point tracker for photovoltaic systems. Solar Energy. 1991;**46**(5):279-282

[2] Zaidi B, Saouane I, Shekhar C. Simulation of single-diode equivalent model of polycrystalline silicon solar cells. International Journal of Materials Science and Applications. 2018;**7**:8-10

[3] Zaidi B, Saouane I, Madhava Rao MV, Li R, Hadjoudja B, Gagui S, et al. Matlab/simulink based simulation of monocrystalline silicon solar cells. International Journal of Materials Science and Applications. 2016;**5**:11-15

[4] Mrabti T, El Ouariachi M, Kassmi K, Olivié F, Bagui F. Amélioration du Fonctionnement des Systèmes Photovoltaïques suite aux Brusques Variations des conditions Météorologiques et de la charge. Revue des Energies Renouvelables. 2008; **11**(1):107-117

[5] Chisti Y. Biodiesel from microalgae. Biotechnology Advances. 2007;**25**(3): 294-306

[6] Zou Z, Ye J, Sayama K, Arakawa H. Direct splitting of water under visible light irradiation with an oxide semiconductor photocatalyst. Nature. 2001;**414**(6864):625-627

[7] Zaidi B, Belghit S, Shekhar C, Hadjoudja B, Chouial B. Impact of anti-reflective coating on the characteristics of a-Si: H solar cells. Nanosistemi Nanomateriali, Nanotehnologii. 2018;**16**:713-718

[8] Pudov AO, Kanevce A, Al-Thani H, Sites JR, Hasoon FS. Secondary barriers in CdS-$CuIn_{1-x}Ga_xSe_2$ solar cells. Journal of Applied Physics. 2005;**97**:1063

[9] Zaidi B, Zouagri M, Merad S, Shekhar C, Hadjoudja B, Chouial B. Boosting electrical performance of CIGS solar cells: Buffer layer effect. Acta Physica Polonica A. 2019;**136**:988-991

[10] Beaucarne G. Silicon thin-film solar cells. Advances in OptoElectronics. 2007;**2007**:1-12

[11] Available from: http://www.photovoltaique.info/Les-types-de-modules

[12] Zaidi B. Introductory chapter: Introduction to photovoltaic effect. In: Solar Panels and Photovoltaic Materials. Rijeka: IntechOpen; 2018

[13] Zaidi B, Belghit S. Silicon Materials. Rijeka: IntechOpen; 2019

[14] Zaidi B. Solar Panels and Photovoltaic Materials. Rijeka: IntechOpen; 2018

Chapter 2

Materials for Photovoltaics: Overview, Generations, Recent Advancements and Future Prospects

Muhammad Aamir Iqbal, Maria Malik, Wajeehah Shahid, Syed Zaheer Ud Din, Nadia Anwar, Mujtaba Ikram and Faryal Idrees

Abstract

As a consequence of rising concern about the impact of fossil fuel-based energy on global warming and climate change, photovoltaic cell technology has advanced significantly in recent years as a sustainable source of energy. To date, photovoltaic cells have been split into four generations, with the first two generations accounting for the majority of the current market. First generation of thin-film technologies is based on monocrystalline or polycrystalline silicon and gallium arsenide cells and includes well-known medium- or low-cost technologies with moderate yields, whereas, second generation includes devices with lower efficiency and manufacturing costs. Third generation is based on novel materials and has a wide range of design options, as well as expensive but highly efficient cells. However, fourth generation, also known as "inorganics-in-organics," combines the low cost and flexibility of polymer thin films with the durability of innovative inorganic nanostructures (metal nanoparticles or metal oxides) in organic-based nanomaterials (carbon nanotubes, graphene, and their derivatives). The aim of this chapter was to highlight the current state of photovoltaic cell technology in terms of manufacturing materials and efficiency by providing a comprehensive overview of the four generations as well as the relevance of graphene and its derivatives in solar cell applications.

Keywords: efficiency, generations, graphene, photovoltaics, polymers

1. Introduction

The rapid growth of the world's community and industrial area is leading to the great need for energy while renewable energy sources consumption such as geothermal, biomass, wind, and photovoltaic sources are challenging the planet's survival having its adverse effects. From the above-listed energy sources, photovoltaics is the technology used for the conversion of sunlight into electrical power by means of semiconductor materials. By considering their history, in 1883, Fritts worked on photovoltaics applications for the first time [1]. In 1954, the p-n junction diode potential was discovered at Bells laboratory with the efficiency of 6% using silicon material [2], and the same work has also been reported to make heterojunction solar cells based on Cu_2S/CdS [3]. These diodes work on the phenomenon of generated voltages when sunlight falls on them. In the 1960s, the photovoltaic system for the first time was employed in commercial applications for space solar cells to deliver the power for satellite applications [4], and silicon semiconductor materials have been reported to be widely used in photovoltaic technology [5]. Moreover, in spite of the extensive use of silicon semiconductor-based technology, it has a high cost, which is the main drawback of not using this technology for home-based device applications. To overcome these challenging problems, researchers have put their efforts to replace silicon-based solar cells technology with the one having superior results [6]. Several researches show numerous classifications of materials, such as organic, inorganic, and hybrid materials, to potentially replace silicon materials from existing solar cells technology [7].

2. Overview of solar cell technology

Solar cells can be categorized according to their material composition whereas silicon-based semiconductors are dominant in the industrial share of photovoltaics, and despite considering the advantages of silicon material in photovoltaics, they lack some factors, such as very low absorbing power as well as needing almost 200–300 semiconducting material films to absorb the incident sunlight.

To overcome the lacking of silicon-based semiconductors, alternative technology was developed with solar cells having a micron-sized thickness, low cost but also low efficiencies [8], while thin sheet solar cells technology was also developed on flexible sheets. The advanced technologies are cadmium telluride, copper indium gallium selenide, and thin film-based technologies with reported efficiency of 20% [9, 10]. Some of the cadmium telluride-based solar cells companies have gained tremendous success at the commercial level with an efficiency of about 11% [11]. However, silicon-based single-junction and multi-junction solar cells have an extensive history of their commercial use including different active layers, such as microcrystalline silicon or hydrogenated amorphous silicon germanium with varying efficiencies from 8 to 12%, but their high cost and low efficiencies have limited the use of silicon material in solar cells and introduced the improved model.

Considering the mechanism of solar cells as shown in **Figure 1**, when sunlight falls on solar cell's surface, it reflects off from the surface. If the photon energy is lesser than materials bandgap energy then incident light is directly passed out from the semiconductor without being absorbed but if the photon energy is more than materials bandgap energy then the semiconductor material absorbs that energy and

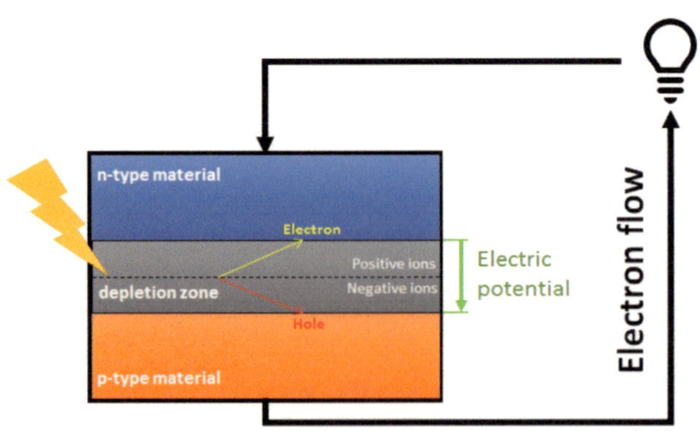

Figure 1.
Schematic diagram of a solar cell [12].

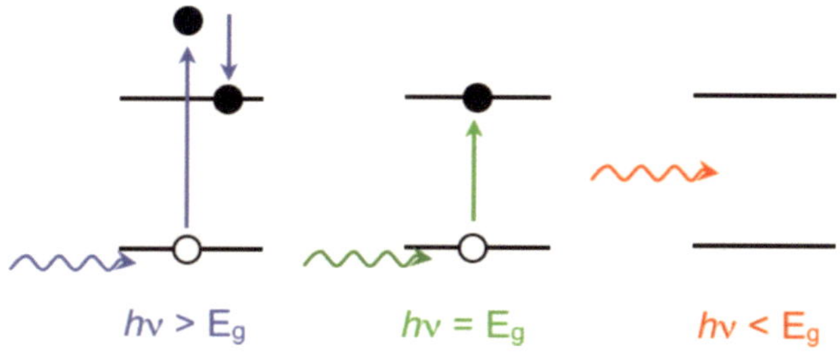

Figure 2.
Photon energy compared to the bandgap energy of a semiconductor [13].

excites electrons from valance to conduction band and generates electron-hole pair as shown in **Figure 2**. The advanced solar cells contain layers, such as the contact layer and window layer, which reduce the reflection losses by increasing the absorption efficiency of solar cells, resulting in electron-hole pairs that are then separated and driven to their collection electrodes. In p-n junction, the charge separation occurs by diffusion method and is collected by electrodes that generate current. However, the recombination process hurdles the carriers to reaching out to the electrodes, therefore not all the charge carriers got captured [13].

2.1 Approaches to light trapping

The solar cell surface is structured to trap the incident light within the semiconductor that enhances absorption over multiple passes while light trapping is the foremost mechanism of advanced solar cells. Silicon-based solar cells have a pyramidal surface structure that allows light to reflect from the cell surface to the silicon layer. The considerations for thick and thin-film light trapping designs are different—for thick films, light trapping can be defined using ray optics, although thin films can be treated with wave optics [13].

2.1.1 Ergodic limit

A thick film light trapping mechanism can be demonstrated with ergodic behavior if the cell texture is fair enough to randomize the light direction within a solar cell. The total internal reflection between the semiconductor material and surrounding medium generates the refractive index contrast which enhances photon path length within the semiconductor material, and hence absorption capacity. However, some portion of light may leave the semiconductor through an exit cone of some angle [13].

2.1.2 Thin films

In the case of smooth thin films, various mechanisms are used to exhibit absorption, such as interface reflection and refraction, material thickness, and light coherence. If the incident light coherency is greater than the film thickness, then it may act as a Fabry–Perot cavity, which serves as a resonant absorber [14]. The Fresnel coefficients are used to calculate reflection and refraction at normal incidence angle with tracked phasors over multi-passes.

3. Solar cell generations

Photovoltaic cells are categorized into four main classes according to their modifications, and these classifications are called generations [15]. **Figure 3** depicts materials that comprise each generation while **Figure 4** represents a historical overview of efficiencies of solar cells.

3.1 First-generation photovoltaic solar cells

The first-generation of photovoltaic solar cells is based on crystalline film technology, such as silicon and GaAs semiconductor materials. Silicon (Si) is the extensively used material for commercial purposes, and almost 90% of the photovoltaic solar cell industry is based on silicon-based materials [17], while GaAs is the oldest material that has been used for solar cells manufacturing owing to its higher efficiency. There are some advantages to use silicon material for photovoltaic solar cells manufacturing, such as:

- It is abundantly found in the earth's crust, which ensures its availability and could be inexpensive [18].
- It is a non-toxic chemical that delays the contamination process and durability loss that may occur when Si is used as a cell material.

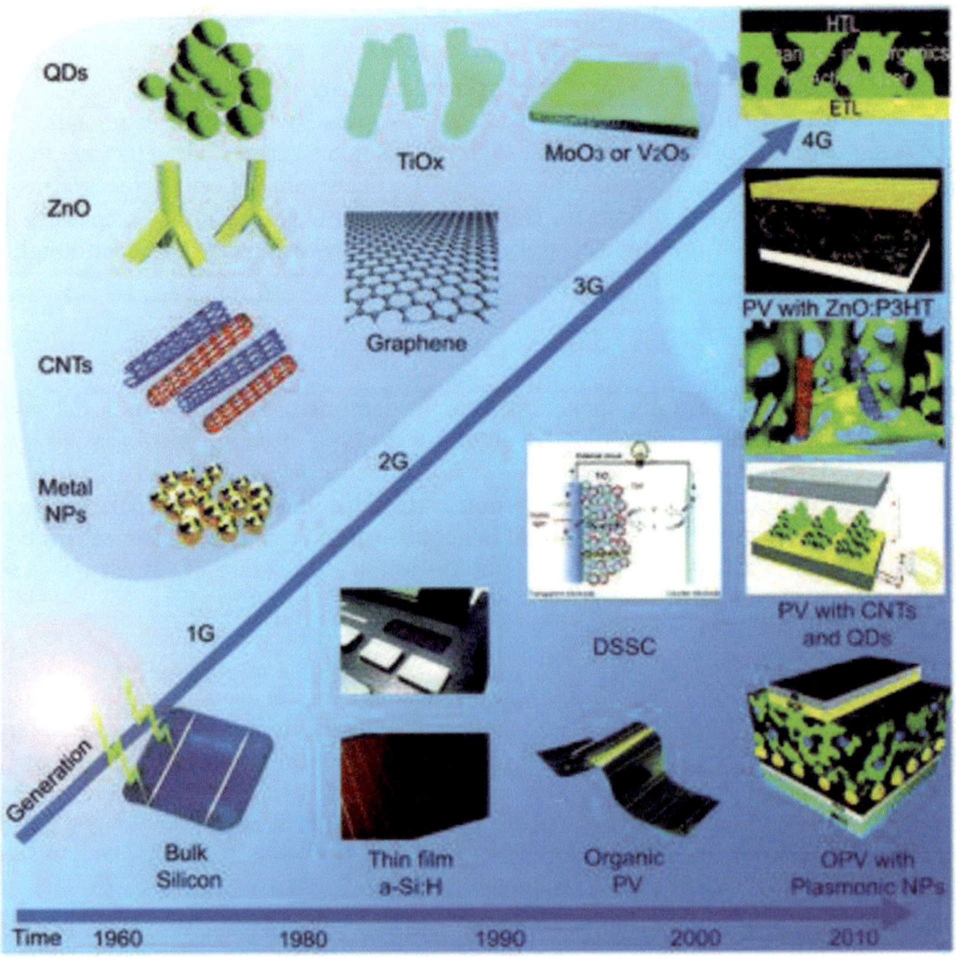

Figure 3.
Four generations of photovoltaic cells along with the materials that comprise each generation [15].

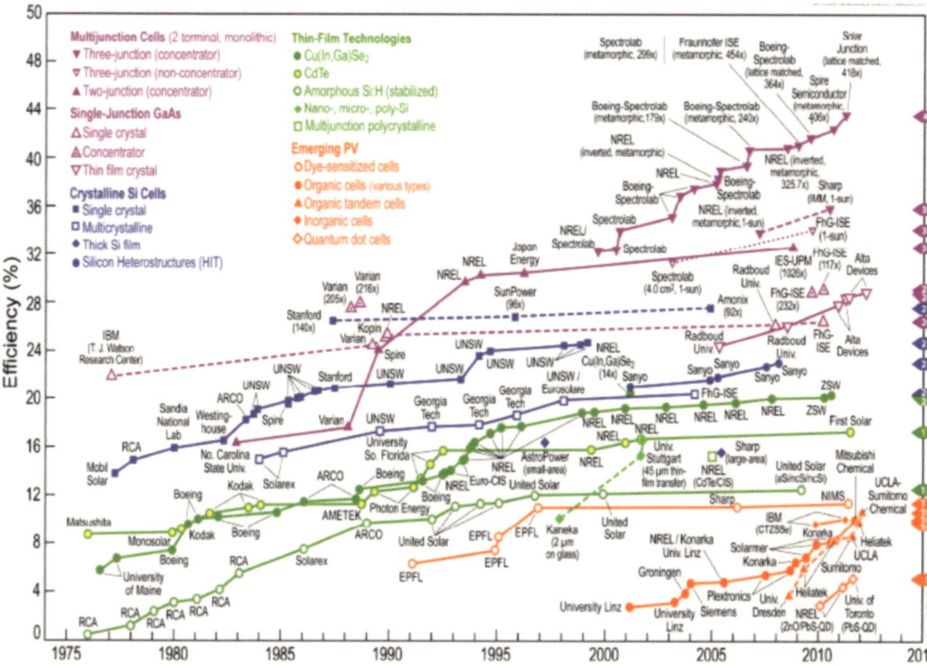

Figure 4.
A historical overview of solar cell efficiency [16].

- Silicon-based photovoltaic solar cells are easily compatible with the silicon-based microelectronic sector, resulting in the creation of the most interesting technologies [17].

The schematic representation of a silicon-based solar cell is shown in **Figure 5**. The GaAs has specialized use in multi-junction cells comprising several semiconductor materials and photovoltaic solar cells [19] having the following special characteristics:

- The GaAs has a bandgap of 1.43 eV that makes it an ideal material for single-junction photovoltaic solar cells.

- It has a good absorption power, so a few micron-thickness cells can produce a very strong absorption spectrum while silicon-based cells need a thick material to reach this limit of absorption.

- Because of the many doping chemicals that can modify the optoelectronic characteristics of a solar cell, GaAs can generate a multifunctional cell structure.

- Also, because of low-temperature coefficients, GaAs-based solar cells have the least influence on temperature.

Hence, the first-generation photovoltaic solar cells can be further classified into monocrystalline, polycrystalline, and GaAs solar cells.

3.2 Second-generation photovoltaic solar cells

The second-generation photovoltaic solar cells have the main focus of cost minimization that was the main issue of first-generation photovoltaic solar cells, and this can be achieved using thin-film technologies by reducing the material quantity as well as

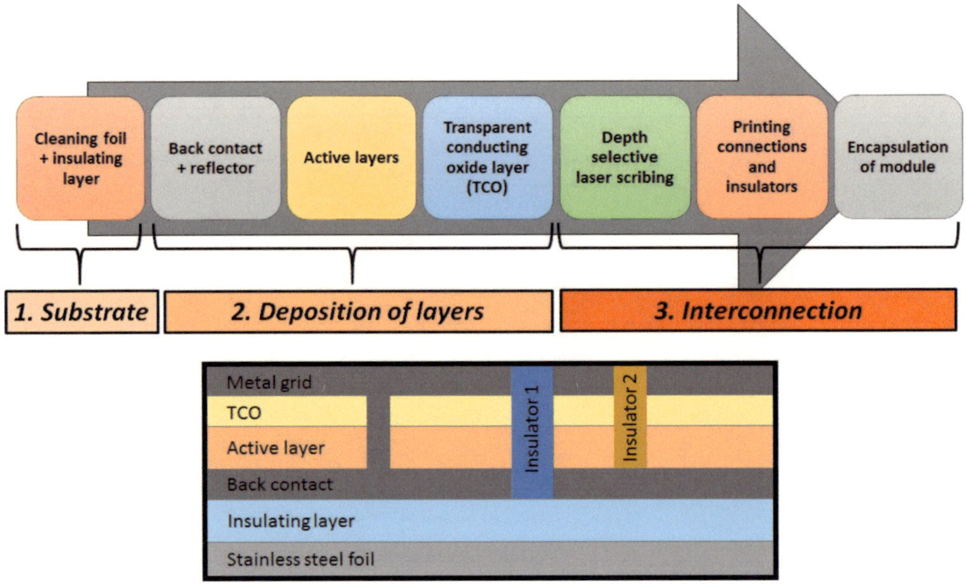

Figure 5.
Schematic representation of a silicon-based solar cell [12].

improving its quality. This modification is based on materials that showed good results in first-generation development and were extended to a-Si, c-Si, CdTe, and copper indium gallium selenide (CIGS) [15]. The advantageous factors of second-generation photovoltaics are listed below [20, 21];

- They are cheaper because of less material usage, up to a few microns in thickness.
- They have a maximum absorption coefficient.
- They can be used for vacuum and non-vacuum processes.
- This technological advancement can reduce the production steps by allowing direct integration to high-voltage modules.

However, some of the drawbacks are still present and listed below:

- Second-generation solar cells are less efficient as only 20% efficiency has been reported for CIGS [16].
- It has a fast degradation phenomenon through induced light, which limits its outdoor applications.
- The availability of raw materials may also be a limiting factor in some technologies.

3.3 Third-generation photovoltaic solar cells

Third-generation photovoltaics emerged from the gap left by second-generation technologies which required improved device efficiency via thin-layer deposition and intend to introduce novel materials with new techniques [22]. This sophisticated technology may be costly but the cost per watt peak would be decreased. As this technique is nontoxic and uses readily accessible materials, it is best suited for large-scale photovoltaic solar cell applications. These materials might be organic or nanostructured with high efficiency of more than 60% achieved by the use of different charge carrier collecting methods [23]. More emphasis has been placed on charge carrier mechanism, charge collecting, and energy capture improvements in third-generation photovoltaic solar cells. The important technologies used in third-generation photovoltaic solar cells are—dye-sensitized solar cells (DSSCs), organic and polymeric solar cells, perovskite cells, quantum dot cells, and multi-junction cells. The considerable advantages of third-generation photovoltaic solar cells may include solution-processable technologies, efficient technologies for commercial production, mechanical toughness, and high efficiencies at higher temperatures. However, the important challenge of this generation is to reduce the cost of solar electricity.

3.4 Fourth-generation photovoltaic solar cells

Fourth-generation photovoltaic solar cells combine the benefits of previous generations, such as lower cost, flexibility, and high stability of third-generation nanomaterials, metal oxides, graphene, and carbon nanotubes. These characteristics will give improved solar cell devices with the needed low-cost manufacturing as well as durability and the usage of nanomaterials in solar devices will aid to increase charge dissociation and transportation inside the cells. Because of its amazing properties and allotropic forms appearing in all four dimensions, graphene is a potential solar cell material with great scientific hopes in fourth-generation technological accomplishments [12].

4. Graphene and graphene-doped solar cells

Given the desire for renewable energy sources and their complete contributions to technological advancement, the demand for solar cells has increased with the added benefit of being low cost and simple to operate. Solar cells are not very efficient in general; it is the substance that makes them so. Recent developments in graphene-based solar cells boosted their efficiency about 20% by lowering the reflection of incident light. To improve the efficiency of solar cells, graphene may be doped in a variety of ways. Here is a thorough review of graphene-based solar cells and their doping variants that are being considered and researched across the world.

4.1 Principles of graphene-based solar cells

Graphene-based solar cells operate in the same way as conventional inorganic solar cells do today but the sole difference is that graphene or graphene-based materials replace inorganic components. Using graphene as a promising material in solar cells improves the adaptability along with tuneability of solar cells, while the number of graphene layers in a device and the doping effects are two essential characteristics that determine the efficiency of graphene-based devices. **Figure 6** depicts the schematic representation for inorganic and organic solar cells with graphene.

4.2 Graphene-silicon solar cells

Solar cells can also be implemented using carbon allotropes that are used for a wide range of applications and made them cost-effective. Some of the carbon allotropes show remarkable results while some are not much efficient because of the inability to tune their electronic properties and layer thickness. Graphene-based solar cells feature such efficient qualities, such as the capacity to modify layer thickness as per the requirement as well as the ability to tune properties based on the material combination. Graphene is used in solar cells with graphene-silicon combination with

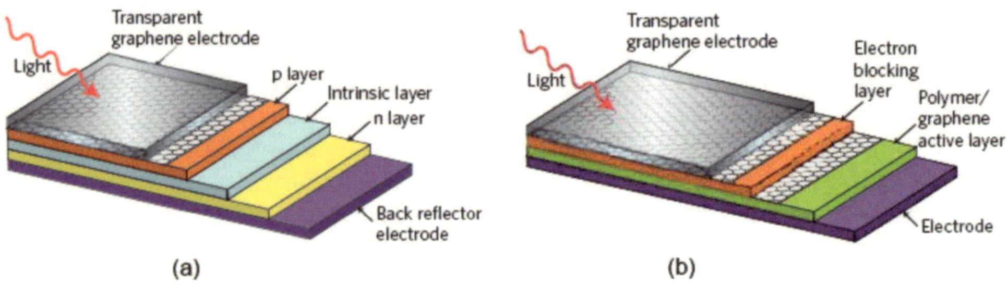

Figure 6.
Schematic representation for (a) inorganic solar cells, and (b) organic solar cells with graphene [24].

both p-type and n-type heterojunctions and as per reported literature, pure silicon solar cell's efficiency is better than any combination. However, graphene tuneability is much better in the case of hybrid solar cells and in the phase of advancement to improve its efficiency compared with pure silicon. According to available research, n-type heterojunctions create about 0.55 internal voltages for electron-hole pair separation while Schottky junctions generate 1.5% efficiency with a filler factor of 56%. As a result, depending on the dopant, efficiency can be increased and gold particles doped in graphene sheets showed a 40% increase in efficiency [25].

4.3 Graphene-polymer solar cells

Graphene-polymer-based solar cells are gaining popularity in the market and these polymer materials may be organic and beneficial due to their simple production method, low cost, and easy tuneability. Graphene has been used in possible applications of coating, layering, and temperature annealing in electrodes as a hybrid material with a combination of organic-inorganic materials. The Fermi level of graphene and semiconductor sheets is closer for charge injection applications and they have strong energy characteristics. Graphene can be combined with polymers to produce a hybrid material with a reported bandgap of 3.6 V that prevents cathode to electrode electron transport. Graphene layers with a thickness of 2 nm are claimed to provide the best outcomes in terms of improved electrical resistance and electron transmission [26].

4.4 Graphene-quantum dot solar cells

Carbon allotropes such as graphene and carbon nanotubes when coupled with quantum dots can produce efficient solar cells and the graphene-quantum dot combination has potential applicable uses. They are synthesized using an electrophoretic and chemical bath deposition method which produces a layered pattern of graphene and quantum dots that have 18 layers of both materials. This combination is said to have 16% efficiency while graphene-carbon nanotubes have 7% efficiency. The combination of graphene and quantum dots provides a superior framework as well as rapid electron transport between graphene and quantum dots while limiting charge recombination [27].

4.5 Graphene-tandem solar cells

Tandem solar cells consist of more than two subcells combined together called multi-junction solar cells. According to studies, a single solar cell is almost 40% efficient but tandem solar cells based on graphene oxides are up to 86% efficient owing to their greater subcell combination, and as a result tandem configuration can improve the energy conversion of solar cells [28].

4.6 Graphene-perovskite solar cells

Perovskite solar cells have an intriguing bandgap that results in remarkable absorption properties which eventually results in increased energy conversion efficiency of a solar cell and because of their standard structure, this type of solar cell can easily be tuned by changing the required material. These cells also feature a graphene layer which accounts for only 0.6 weight-percent of a cell and any other proportion reduces device efficiency. This cell arrangement is a little more complicated but it results in higher energy conversion efficiency when compared to non-graphene oxide layer solar cells. When graphene oxide materials are employed in this sort of solar cell and coupled with a light absorber they serve as a hole-conductor. Considering graphene oxide's other applications, it increases interface wettability on the surface of a perovskite solar cell and reduces the hole transporting layer contact angle. In addition, the C-C bonds of graphene sheets absorb the hole transport layer molecules which improve interfacial contacts of solar cells and results in better device performance [29].

4.7 Graphene-organic solar cells

Organic solar cells like inorganic ones play an important role in industrial applications and because of their endurance and low cost, organic devices are becoming more popular in technology. Major inorganic components in solar cells are being replaced by hybrid organic-inorganic components that exhibit superior physical and chemical characteristics as well as easy processability, easy availability, environmental friendliness, and most significantly, cost-effective synthesis and manufacturing. The major disadvantage of recent solar cells is their environmental impact; however organic components address this issue by offering stability against chemical deterioration, temperature, humidity, and other factors [30].

4.8 Graphene bulk-heterojunction solar cells

Graphene is a potential material used in various applications including electrodes, donor-acceptor layers as well as active layers because of its promising properties, such as conductivity, flexibility, and transparency, that make it an efficient material. Multi-junction solar cells also depend on graphene specificity such as its thickness, annealing temperature, doping concentration, and so on. Graphene heterojunctions employ graphene-based solar cells as well and they all rely on how graphene and its derivatives are mixed in these cells. Transparent electrodes, gallium arsenide (GaAs) solar cells, and photovoltaic layers are the primary components of graphene heterojunction solar cells [31].

5. Future prospects

The photovoltaic solar cell is a fascinating research area in both academic and industrial fields with advances and breakthroughs occurring on a regular basis. Solar cells [32] are often manufactured utilizing silicon and inorganic materials which have significant limits in current uses. The development of improved organic graphene-based and graphene derivative-based materials, as well as narrow bandgap polymers, is revolutionizing the market by demonstrating their potential and perfect characteristics necessary for solar cell device structure applications. There has been tremendous development in the improvement of graphene-based solar cells and more work has to be done in the future to achieve further growth in this industry. Because of their capacity to enhance different characteristics, graphene-based solar cells are customizable and adaptable to future limits in solar research. However, by enhancing existing solar cells and illuminating the characteristics of already employed non-graphene-based solar cells or developing a new variety of graphene photovoltaics, it is clear that graphene will play an important role in this intriguing analog [33].

6. Concluding remarks

An overview of photovoltaic solar cells is provided in this chapter, along with illustrations of four generations as well as prospective applications of graphene and graphene-based materials. Briefly, the first-generation thin-film technology was based on monocrystalline or polycrystalline silicon cells and gallium arsenide while the second generation includes devices with lower efficiency and lower manufacturing costs. The third generation has novel materials and a wide range of design options and expensive but highly efficient cells; however, the fourth generation, also known as "inorganics-in-organics," combines the low cost/flexibility of polymer thin films with the durability of innovative inorganic nanostructures in organic-based nanomaterials. Recent developments in graphene-based solar cells boosted their efficiency about 20% by lowering the reflection of incident light.

Author details

Muhammad Aamir Iqbal[1*], Maria Malik[2], Wajeehah Shahid[3], Syed Zaheer Ud Din[4], Nadia Anwar[3,5], Mujtaba Ikram[6] and Faryal Idrees[7]

1 School of Materials Science and Engineering, Zhejiang University, Hangzhou, China

2 Centre of Excellence in Solid State Physics, University of the Punjab, Lahore, Pakistan

3 Department of Physics, The University of Lahore, Lahore, Pakistan

4 International School of Optoelectronic Engineering, Qilu University of Technology (Shandong Academy of Sciences), Jinan, China

5 School of Materials Science and Engineering, Tsinghua University, Beijing, China

6 Institute of Chemical Engineering and Technology, University of the Punjab, Lahore, Pakistan

7 Department of Physics, University of the Punjab, Lahore, Pakistan

*Address all correspondence to: maamir@zju.edu.cn

References

[1] Mirica MC, Tudoran MA, Putz MV. Photovoltaic system. In: New Frontiers in Nanochemistry. Palm Bay, FL: Apple Academic Press; 2020. pp. 347-358

[2] Chapin DM, Fuller CS, Pearson GL. A new silicon pn junction photocell for converting solar radiation into electrical power. In: Semiconductor Devices: Pioneering Papers. Singapore: World Scientific; 1991. pp. 969-970

[3] Reynolds DC, Leies G, Antes LL, Marburger RE. Photovoltaic effect in cadmium sulfide. Physical Review. 1954; **96**(2):533

[4] Hegedus S, Luque A. Achievements and challenges of solar electricity from photovoltaics. In: Handbook of Photovoltaic Science and Engineering. UK: John Wiley & Sons Ltd; 2011. pp. 1-38

[5] Kong L, Wang J, Luo T, Meng F, Chen X, Li M, et al. Novel pyrenehexafluoroisopropanol derivative-decorated single-walled carbon nanotubes for detection of nerve agents by strong hydrogen-bonding interaction. The Analyst. 2010;**135**(2): 368-374

[6] Nayak PK, Mahesh S, Snaith HJ, Cahen D. Photovoltaic solar cell technologies: Analysing the state of the art. Nature Reviews Materials. 2019; **4**(4):269-285

[7] Almeida MA. Recent advances in solar cells. Solar Cells. 1st ed. Switzerland: Springer; 2020:79-122. DOI: 10.1007/978-3-030-36354-3_4. eBook ISBN 978-3-030-36354-3

[8] Green MA. Third generation photovoltaics: Solar cells for 2020 and beyond. Physica E: Low-Dimensional Systems and Nanostructures. 2002; **14**(1-2):65-70

[9] Romeo A, Terheggen M, Abou-Ras D, Bätzner DL, Haug FJ, Kälin M, et al. Development of thin-film Cu (In, Ga) Se2 and CdTe solar cells. Progress in Photovoltaics: Research and Applications. 2004;**12**(2-3):93-111

[10] Contreras MA, Egaas B, Ramanathan K, Hiltner J, Swartzlander A, Hasoon F, et al. Progress toward 20% efficiency in Cu (In, Ga) Se2 polycrystalline thin-film solar cells. Progress in Photovoltaics: Research and Applications. 1999; **7**(4):311-316

[11] Schmidtke J. Commercial status of thin-film photovoltaic devices and materials. Optics Express. 2010; **18**(103):A477-A486

[12] Luceno-Sanchez JA, Díez-Pascual AM, Peña CR. Materials for photovoltaics: State of art and recent developments. International Journal of Molecular Sciences. 2019;**20**(4):976

[13] Ferry VE, Verschuuren MA, Li HB, Verhagen E, Walters RJ, Schropp RE, et al. Light trapping in ultrathin plasmonic solar cells. Optics Express. 2010;**18**(102):A237-A245

[14] Heavens OS. Optical properties of thin solid films. North Chelmsford, MA: Courier Corporation; 1991

[15] Jayawardena KI, Rozanski LJ, Mills CA, Beliatis MJ, Nismy NA, Silva SR. 'Inorganics-in-Organics': Recent developments and outlook for 4G polymer solar cells. Nanoscale. 2013;**5**(18):8411-8427

[16] Blakers A, Zin N, McIntosh KR, Fong K. High efficiency silicon solar cells. Energy Procedia. 2013;**33**:1-10

[17] Sampaio PG, González MO. Photovoltaic solar energy: Conceptual framework. Renewable and Sustainable Energy Reviews. 2017;**74**:590-601

[18] Kuhlmann AM. The second most abundant element in the earth's crust. Journal of Metals. 1963;**15**(7):502-505

[19] Moon S, Kim K, Kim Y, Heo J, Lee J. Highly efficient single-junction GaAs thin-film solar cell on flexible substrate. Scientific Reports. 2016;**6**(1):1-6

[20] Grandidier J, Callahan DM, Munday JN, Atwater HA. Gallium arsenide solar cell absorption enhancement using whispering gallery modes of dielectric nanospheres. IEEE Journal of Photovoltaics. 2012;**2**(2): 123-128

[21] Choubey PC, Oudhia A, Dewangan R. A review: Solar cell current scenario and future trends. Recent Research in Science and Technology. 2012;**4**:99-101

[22] Conibeer G. Third-generation photovoltaics. Materials Today. 2007;**10**(11):42-50

[23] Chopra KL, Paulson PD, Dutta V. Thin-film solar cells: An overview. Progress in Photovoltaics: Research and applications. 2004;**12**(2-3):69-92

[24] Bonaccorso F, Sun Z, Hasan T, Ferrari AC. Graphene photonics and optoelectronics. Nature Photonics. 2010;**4**(9):611-622

[25] Won R. Graphene–silicon solar cells. Nature Photonics. 2010;**4**(7):411

[26] Lin XF, Zhang ZY, Yuan ZK, Li J, Xiao XF, Hong W, et al. Graphene-based materials for polymer solar cells. Chinese Chemical Letters. 2016;**27**(8): 1259-1270

[27] Guo CX, Yang HB, Sheng ZM, Lu ZS, Song QL, Li CM. Layered graphene/quantum dots for photovoltaic devices. Angewandte Chemie International Edition. 2010;**49**(17): 3014-3017

[28] Sinha S, Swain BP. Graphene for the potential renewable energy applications. In: Advances in Greener Energy Technologies. Singapore: Springer; 2020. pp. 439-450

[29] Agresti A, Pescetelli S, Taheri B, Del Rio Castillo AE, Cinà L, Bonaccorso F, et al. Graphene–perovskite solar cells exceed 18% efficiency: a stability study. ChemSusChem. 2016;**9**(18):2609-2619

[30] Pan Z, Gu H, Wu MT, Li Y, Chen Y. Graphene-based functional materials for organic solar cells. Optical Materials Express. 2012;**2**(6):814-824

[31] Singh E, Nalwa HS. Graphene-based bulk-heterojunction solar cells: A review. Journal of Nanoscience and Nanotechnology. 2015;**15**(9):6237-6278

[32] Zaidi B. Introductory chapter: Introduction to photovoltaic effect. Solar Panels and Photovoltaic Materials. UK: IntechOpen; 2018:1-8. Available from: https://www.intechopen.com/books/6691

[33] Wolf EL. Applications of graphene: an overview. Berlin, Germany: Springer Science & Business Media; 2014

Chapter 3

Binary Semiconductors Thin Films Characterization for Solar Cells Applications

Kenza Kamli and Zakaria Hadef

Abstract

The increasing development of technologies based on the thin films, imposed a high quality of these films. The crucial importance for all applications of thin films is related to the stability of their physical and morphological properties. Therefore, to optimize the performances of the thin films it is recommended to study carefully all their parameters in order to enhance the elaborated films. With this intention, various characterizations methods were developed and carried out to study the different qualities of thin films. In this chapter, we take an interest to the study of the characteristics of some binary semiconductors thin films elaborated by ultrasonic spray pyrolysis, and which are destined for solar cells applications. Several used characterizations techniques to the determination of the thin films properties will be given; namely: X-rays diffraction (XRD), Scanning Electron Microscopy (SEM), EDS (Energy Dispersive Spectroscopy), Hall effect and spectrophotometry will be discussed in detail.

Keywords: Thin films, ultrasonic spray, X-rays diffraction, SEM, EDS, spectrophotometry, Hall effect, Binary semiconductors

1. Introduction

The materials have different properties, which are described by their structure, morphology and chemical composition. The determination of these properties and the study of the different characteristics of the given materials in order to their development is very necessary to achieve the requirements of the new technologies. Therefore, materials characterization is a fundamental process in the field of materials science.

While many characterization techniques have been practiced for centuries, such as basic optical microscopy, but new techniques and methodologies are constantly emerging. Detection ranges of the wide variety of instrumental analytical techniques can be summarized versus the probe sizes/resolution as shown in **Figure 1**.

This chapter is devoted to describe certain important characterization techniques used in general to study and develop the different characteristics of certain binary semiconductors thin films, which are used in solar cells, to get layers with high performances.

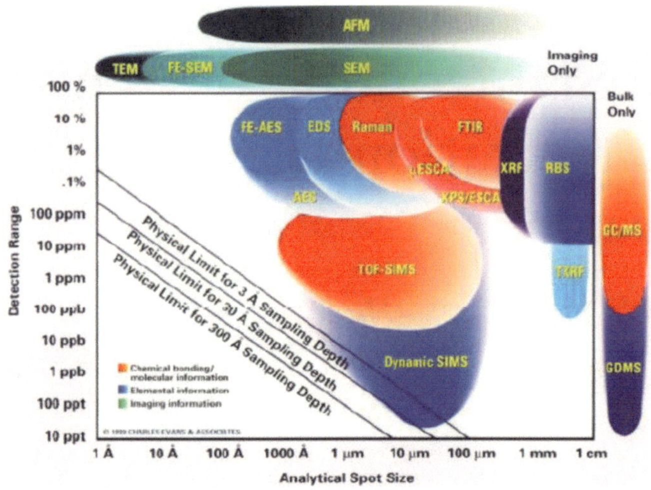

Figure 1.
Schematic of detection ranges versus probe sizes/resolution.

2. X-ray diffraction technique

The use of X-ray methods in the field of materials analysis is now entering its eighth decade. X-ray diffraction techniques are a very useful characterization tool to study, non-destructively, the crystallographic structure, chemical composition and physical properties of materials and thin films. It can also be used to measure various structural properties of these crystalline phases such as strain, grain size, phase composition, and defect structure.

2.1 X-rays creation

The processes of X-rays creation are based on the energy loss of the energetic electrons. **Figure 2a** shows the process of elastic and inelastic scattering where the deflection, or more specifically the acceleration during the deflection would always produce radiation.

Two routes including energy transfer between the incident electron and the electrons of the atom, exists. Both of these processes involve a primary ionization where a core electron is ejected from the atom. The ejected electron falls with an excess energy, which can be disposed as an X-ray or Auger electron [1]. The characteristic X-ray carries the full energy difference of the two electron states as shown in **Figure 2b**. Furthermore, an X-ray for a diffraction experiment is characterized by its wavelength, λ, but the energy, E, is typically more useful.

For XRD analysis, it is always required to use a coherent beam of monochromatic X-rays with a known wavelength [2]. That is why a right selection of metal anode and energy (i.e., a known wavelength) of accelerated electrons is very necessary.

2.2 Principle of measurements of X-rays diffraction

A crystal lattice consists of a regular arrangement of atoms, with layers of high atomic density existing throughout the crystal structure. Knowledge of how atoms are arranged into crystal structures and microstructures is the foundation on which we build our understanding of the synthesis, structure and properties of materials [1].

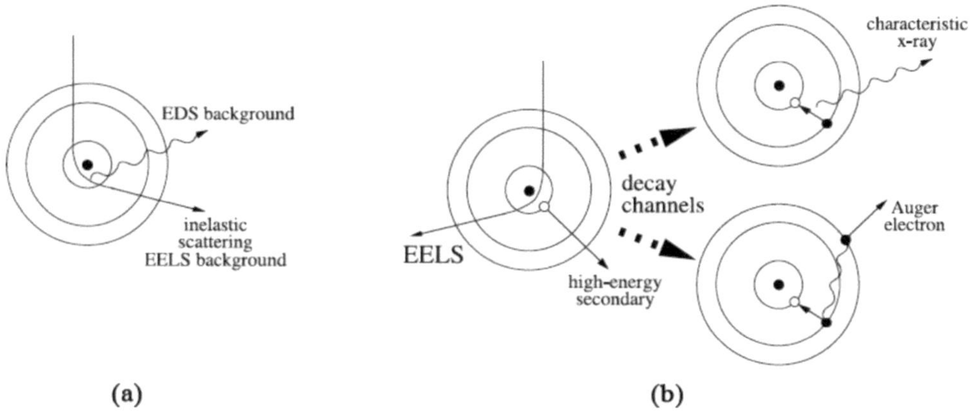

Figure 2.
Interaction processes between a high-energy electron and an atom.

Planes of high atomic density means planes of high electron density. W.L. Bragg derived a simple equation treating diffraction as reflection from planes in the lattice.

$$n\lambda = 2d_{hkl}\sin\theta \quad (1)$$

where n is an integer, d_{hkl} is the distance inter reticular separating the plans defined by the indices from Miller (h, k, l), θ the angle of incidence and thus of reflexion compared to these plans and finally λ is the wavelength of photons X.

In fact, d_{hkl} does not represent the inter-reticular distance only, but it is defined too:

- the vector drawn from the origin of the unit cell to intersect the crystallographic plane (hkl) at a 90° angle, called also the vector magnitude;
- and, it is a geometric function of the size and shape of the unit cell.

Bragg's Law express and interpret the interaction of X-rays with sample, which creates secondary "diffracted" beams generated in the form of cones of X-rays. These beams are related to interplanar spacings in the crystalline powder according to the Bragg's mathematical relation.

Consequently, a family of planes produces a diffraction peak only at a specific angle 2θ. Diffraction experiments are generally made at a fixed wavelength (as we have mentioned above), thus a measure of the diffraction angles will allow the associated d-spacings to be calculated.

Figure 3 show the Bragg X-ray diffraction condition.

X-ray diffraction peaks are produced by constructive interference of monochromatic beam scattered from each set of lattice planes at specific angles. X-ray diffraction from crystalline solids occurs as a result of the interaction of X-rays with the electron charge distribution in the crystal lattice. The ordered nature of the electron charge distribution, which is distributed around atomic nuclei and is regularly arranged with translational periodicity, means that superposition of the scattered X-ray amplitudes will give rise to regions of constructive and destructive interference producing a diffraction pattern.

Bragg's equation tells us about the position of the diffraction peaks (in terms of θ), but tells us nothing about the intensity. The intensities of the diffraction peaks are determined by the arrangement of atoms in the entire crystal. These intensities can

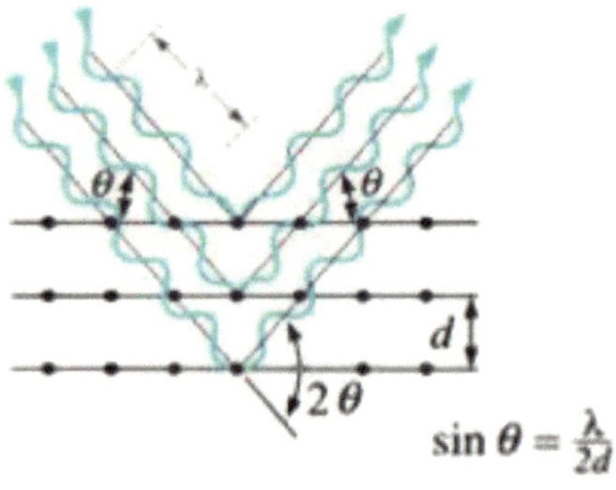

Figure 3.
Bragg X-ray diffraction condition [3].

be explained by the variation of the square of the structure factor according to the following equation [4]:

$$I_{hkl} \alpha |F_{hkl}|^2 / F_{hkl} = \sum_{j=1}^{M} N_j f_j \exp\left[2\pi i\left(hx_j + ky_j + lz_j\right)\right] \quad (2)$$

This factor, F_{hkl}, represent the sums of resulting scattering from all of the atoms in the unit cell to form a diffraction peak from the (hkl) planes of atoms.

N_j is the fraction of every equivalent position that is occupied by atom j.

The three factors: x_j, y_j and z_j, which are the fractional coordinates, represents the atoms position in the atomic planes, and gives the first information about the amplitude of scattered light.

The other information of the amplitude of scattered light is given by the scattering factor f_j, which quantifies the efficiency of X-ray scattering at any angle by the group of electrons in each atom.

2.3 Sample preparation and diffractometer

Sample preparation is usually the most critical factor influencing the quality of the analytical data. Preferably, the sample should exhibit a plane or flattened surface.

All conventional X-ray spectrometers comprise three basic parts: the primary source unit, the spectrometer itself and the measuring electronics. The acquisitions are generally carried out using a goniometer θ-2θ and by using a linear detector.

The diffraction pattern is collected by varying the incidence angle of the incoming X-ray beam by θ and the scattering angle by 2θ while measuring the scattered intensity I(2θ) as a function of the latter. Wide number of powder samples have been measured by using these tools, but it is also applied to the investigation of thin films.

Nowadays, CCD detector or scintillation are used in the novel generation of X-ray Diffractometer detector to record the angles and intensities of the diffracted beams with high resolution.

2.4 X-ray diffraction applications

X-ray Diffraction is considered as one of the most useful characterization techniques, because it is capable of providing general purpose qualitative and quantitative information on the presence of phases in an unknown mixture. This technique uses X-ray (or neutron) diffraction on powder or microcrystalline samples, where ideally every possible crystalline orientation is represented equally.

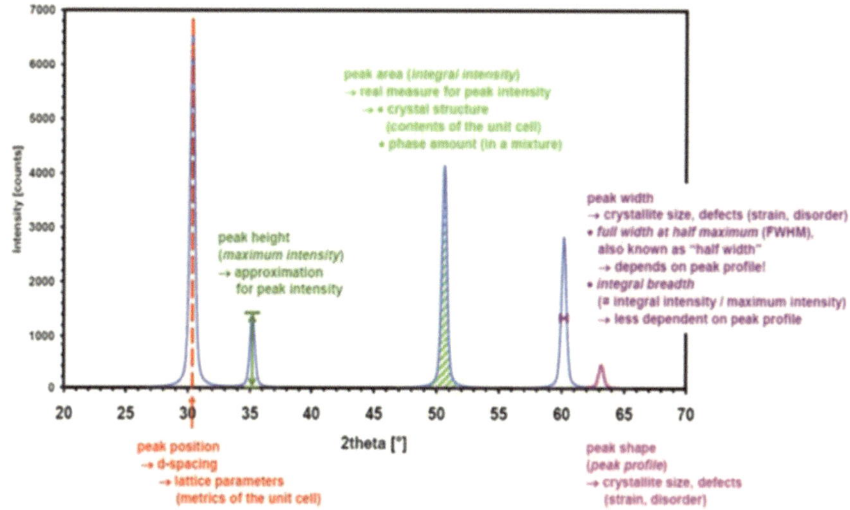

Figure 4.
Information content of an idealized diffraction pattern [5].

Figure 4 shows the information we can get from an idealized diffraction pattern.

A diffraction pattern contains a good deal of information of which three parameters are of special interest:

- The position of the diffraction maxima;
- The peak intensities;
- The intensity distribution as a function of diffraction angle.

The three pieces of information can be used in particular to identify and quantify the contents of the sample, as well as to calculate the material's crystallite size and distribution, crystallinity, stress and strain.

The most traditional use of XRD can be summarized as shown below.

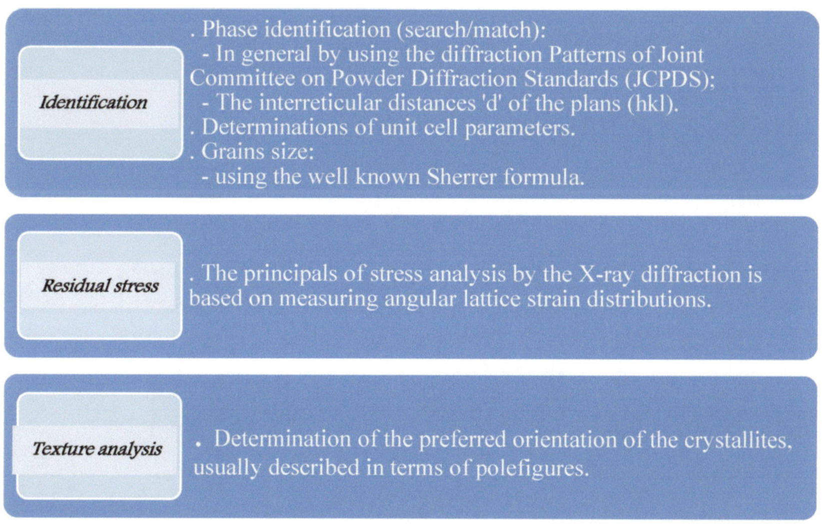

The diffraction pattern is like a fingerprint of the crystal structure. It is a powerful and rapid technique for identification of an unknown material.

2.4.1 Phases identification

Among these applications, the identification of unknown crystals in a sample seems to be the most important. The idea is to match the positions and the intensities of the peaks in the observed diffraction pattern to a known pattern of peaks from a standard sample or from a calculation.

A single crystal specimen in a diffractometer would produce only one family of peaks in the diffraction pattern as given in **Figure 5**. According to this figure, we can resume the most common cases of diffraction peaks identification used in reality.

From **Figure 5**, we can distinguish three cases:

- The first case for **Figure 5a** shows at 2θ = 20.6°, that Bragg's law fulfilled for the (100) planes, producing a diffraction peak;

- The second for **Figure 5b**, the (110) planes would diffract at 2θ = 29.3°; however, they are not properly aligned to produce a diffraction peak (the perpendicular to those planes does not bisect the incident and diffracted beams). Only background is observed;

- And the last one for **Figure 5c**, demonstrate that the (200) planes are parallel to the (100) planes. Therefore, they also diffract for this crystal. Since d_{200} is ½ d_{100}, they appear at 2θ = 42°.

Nevertheless, a polycrystalline sample should contain thousands of crystallites as given in **Figure 5d**. Therefore, all possible diffraction peaks should be observed. For every set of planes, there will be a small percentage of crystallites that are properly oriented to diffract (the plane perpendicular bisects the incident and diffracted beams).

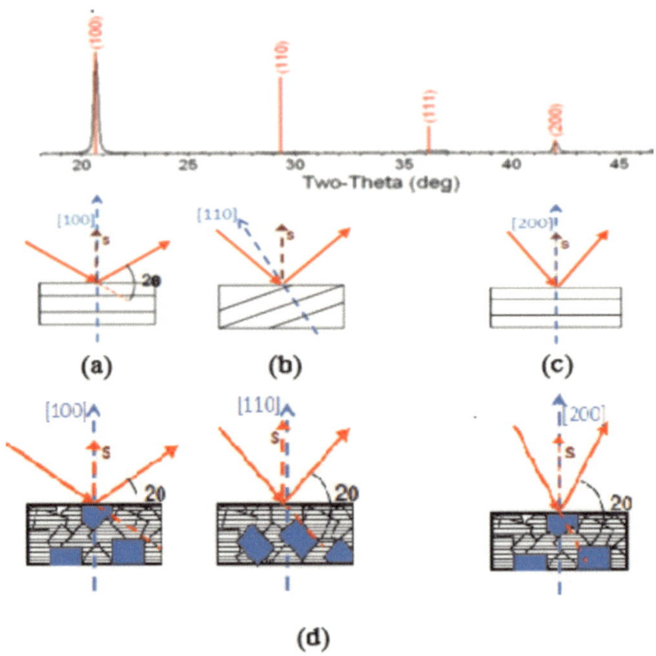

Figure 5.
Information content of an idealized diffraction pattern [5].

In addition, the position and the form of the peaks could give us some important information about the lattice stresses.

As shown in **Figure 6**, it can be clearly noticed the difference between the stressed and none stressed peaks. Cause, in the case of a uniform strain we note a peak move with no shape changes, contrary to the non-uniform strain which produce with the peak displacement a peak broadens.

2.4.2 Grain size

The determination of the crystallite size is based on the calculation of the width with middle height of peak (Full Width at Half Maximum FWHM) expressed in radian, according as mentioned before, to the well-known Sherrer formula (**Figure 7**).

The identification of this parameter is as follow:

An increase in the width of the diffraction causes a decrease in the crystallite size. Destructive interference of the all diffracted beams will result a sharp peak. Complete destructive interference beside the Bragg angle is produced from taking diffraction of large number of planes. Thereby, broadened peak can be observed when the crystallites are of very small sizes in which there are not enough planes to produce complete destructive interference.

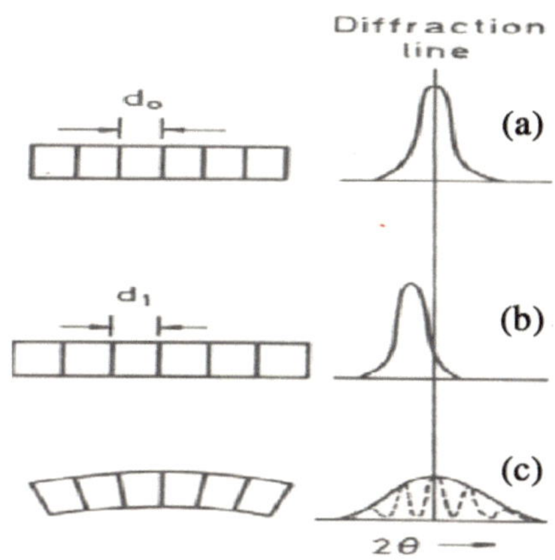

Figure 6.
Strain effect on the diffraction peak [6].

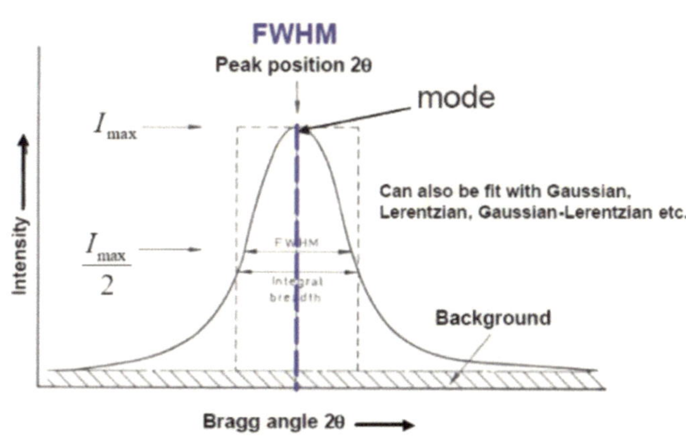

Figure 7.
Determination of the FWHM [6].

3. Scanning electron microscopy

Since its first commercialization, Scanning Electron Microscope (SEM) shows a remarkable progress. This technique is now well known and used in many laboratories. A SEM can be utilized for high magnification imaging of almost all materials.

3.1 Principle

The SEM instrument is very suitable for different kinds of investigations. It is possible to investigate for example the fiber in wood and paper, metal fracture surfaces, production defects in rubber and plastic ect. [7]. Therefore, to be able to interpret the different images and information's collected by using SEM it is essential to understand the principal of function of this tool.

The principle of operation is as follows:

- In a vacuum enclosure, an electron beam focused sweeps the surface of the sample. According to the physicochemical nature of surface, secondary electrons, retrodiffused or Auger electrons are emitted, certain electrons are transmitted and others still dennent place has cathode-luminescence and x-rays as shown in **Figure 8**.

- According to the type of detector used, the retro diffused electrons provide an image topographic (contrast function of the relief) or an image of composition (contrast function atomic number).

For the observation of the PZT, a light metallization is necessary so to evacuate the loads.

3.2 Description of the microscope

A knowledgeable SEM operator should have a basic of contents, which can simplify at this point:

- A source (electron gun) of the electron beam which is accelerated down the column;

- A series of lenses (condenser and objective) which act to control the diameter of the beam as well as to focus the beam on the specimen;

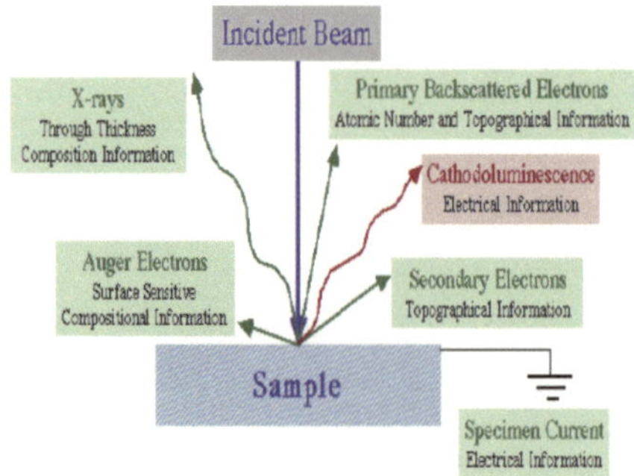

Figure 8.
Principe of electron beam/specimen interactions.

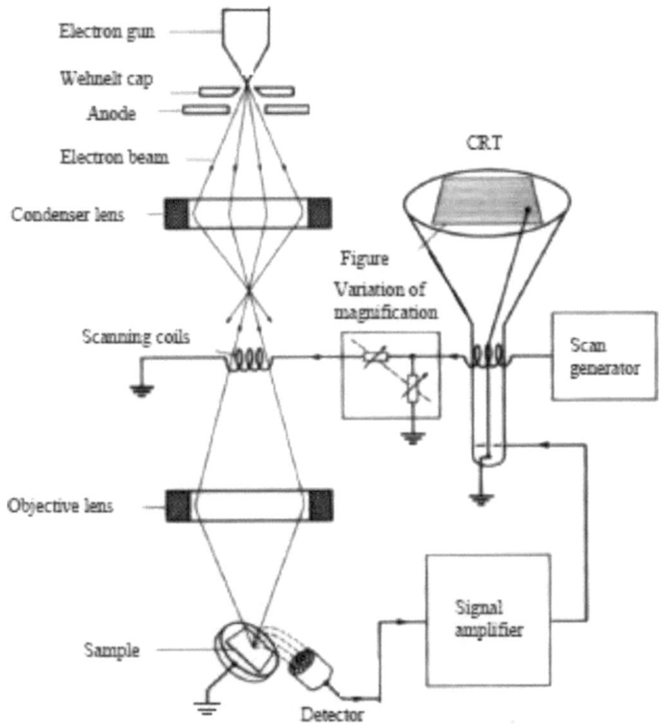

Figure 9.
Scheme of scanning electron microscope [8].

- A series of apertures (micron-scale holes in metal film) which the beam passes through and which affect properties of that beam;
- Controls for specimen position (x, y, z-height) and orientation (tilt, rotation);
- An area of beam/specimen interaction that generates several types of signals that can be detected and processed to produce an image or spectra;
- System of pumping: 3 ionic pumps (vacuum of the column), a diffusion pump equipped with a vane pump (vacuum of the room) and a vane pump for hopper (**Figure 9**).

3.3 SEM applications

The different applications of this technique can be summarized as follow:

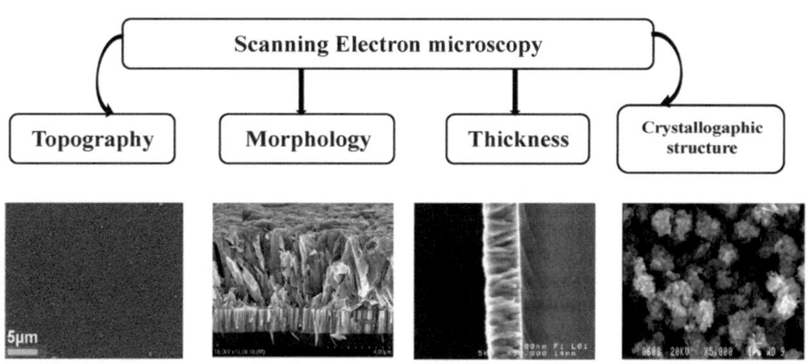

4. Energy dispersive spectroscopy EDS

Energy dispersive X-ray spectroscopy (EDS, EDX or EDXRF) is an analytical technique used for the elemental analysis or chemical characterization of a sample (**Figure 10**).

4.1 Principle

The operating principle of EDS is based on the interaction between electromagnetic radiation and matter; the X-rays emitted by the matter in response will be collected and analyzed. This analysis is due to the fact that each element has its own specific atomic structure; this latter allows the characteristic X-rays of the element atomic structure to be identified uniquely from each other (**Figure 11**) [9].

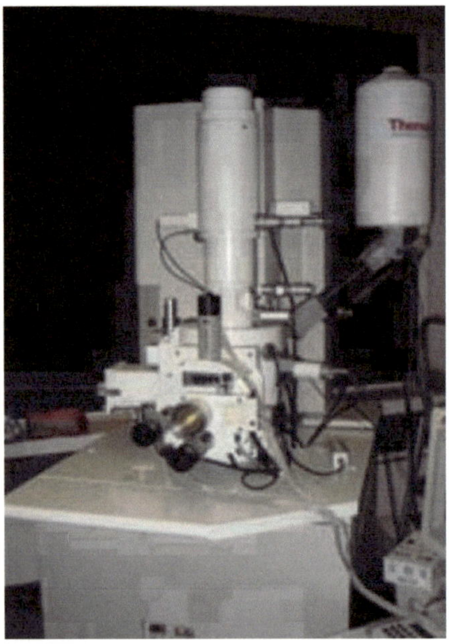

Figure 10.
Energy dispersive spectroscopy (EDS).

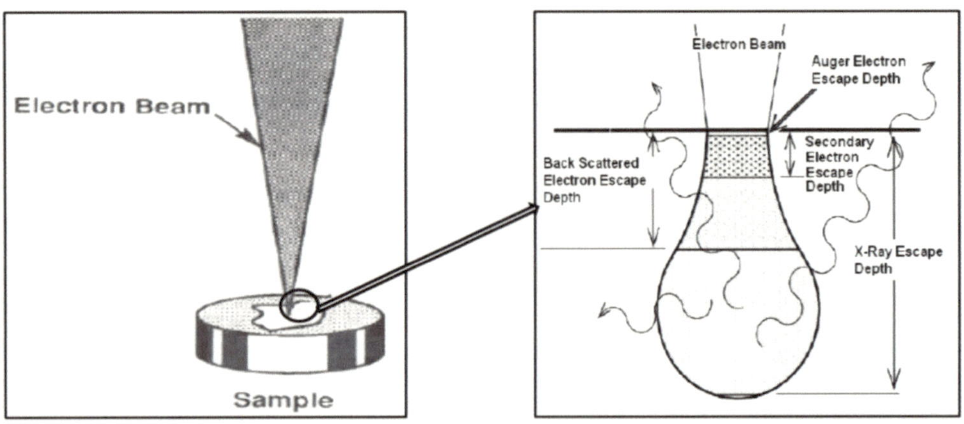

Figure 11.
X-ray source region, with path of X-rays to the spectrometer.

High energy beam of charged particles is focused into the sample being studied in order to stimulate the emission of characteristic X-rays from the specimen. Furthermore, an Energy Dispersive Spectrometer, EDS, is used in order to measure the number and energy of the X-rays emitted from a specimen. As the energy of the X-rays are characteristic of the difference in energy between the two shells, and of the atomic structure of the element from which they were emitted, this allows the elemental composition of the specimen to be measured [9].

4.2 EDS applications

The Energy-dispersive spectrometer is especially useful for qualitative analysis because a complete spectrum can be obtained very quickly, but is also used for the Quantitative analysis. This method permit to find what elements are present and their quantities in an 'unknown' specimen by identifying the lines in the X-ray spectrum using tables of energies or wavelengths (**Figure 12**) [10].

Six types of major artifacts may possibly be generated during the detecting process:

- Peak Broadening;
- Peak distortion;
- Silicon x-ray escape peaks;
- Sum peaks;
- Silicon and gold absorption edges;
- Silicon internal fluorescence peak.

5. Spectrophotometry

The spectrophotometer has well been called the workhorse of the modern laboratory. In particular, ultraviolet and visible spectrophotometry is the method of choice in most laboratories concerned with the identification and measurement of organic and inorganic compounds in a wide range of products and processes.

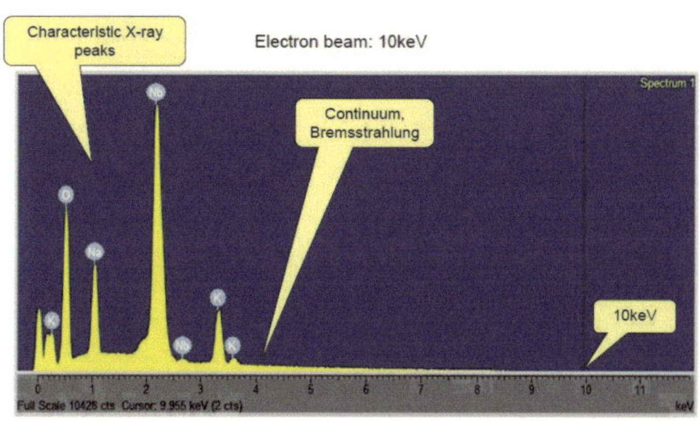

Figure 12.
EDX spectrum of (K,Na)NbO$_3$.

5.1 Principle

Spectrophotometry is commonly divided in two spectroscopic analyses:

- IR spectroscopy (Infrared spectroscopy), deals with the infrared region of the electromagnetic spectrum, that is light with a longer wavelength and lower frequency than visible light. It covers a range of techniques, mostly based on absorption spectroscopy. As with all spectroscopic techniques, it can be used to identify and study chemicals. The infrared spectrum of a sample is recorded by passing a beam of infrared light through the sample. When the frequency of the IR is the same as the vibrational frequency of a bond, absorption occurs.

- Ultraviolet–visible spectroscopy or ultraviolet–visible spectrophotometry (UV–Vis or UV/Vis) refers to absorption spectroscopy or reflectance spectroscopy in the ultraviolet–visible spectral region. This means it uses light in the visible and adjacent (near-UV and near-infrared (NIR)) ranges (**Figure 13**).

5.2 Description of the spectrophotometer

The instrument used in ultraviolet–visible spectroscopy is called a UV/Vis spectrophotometer (**Figure 14**).

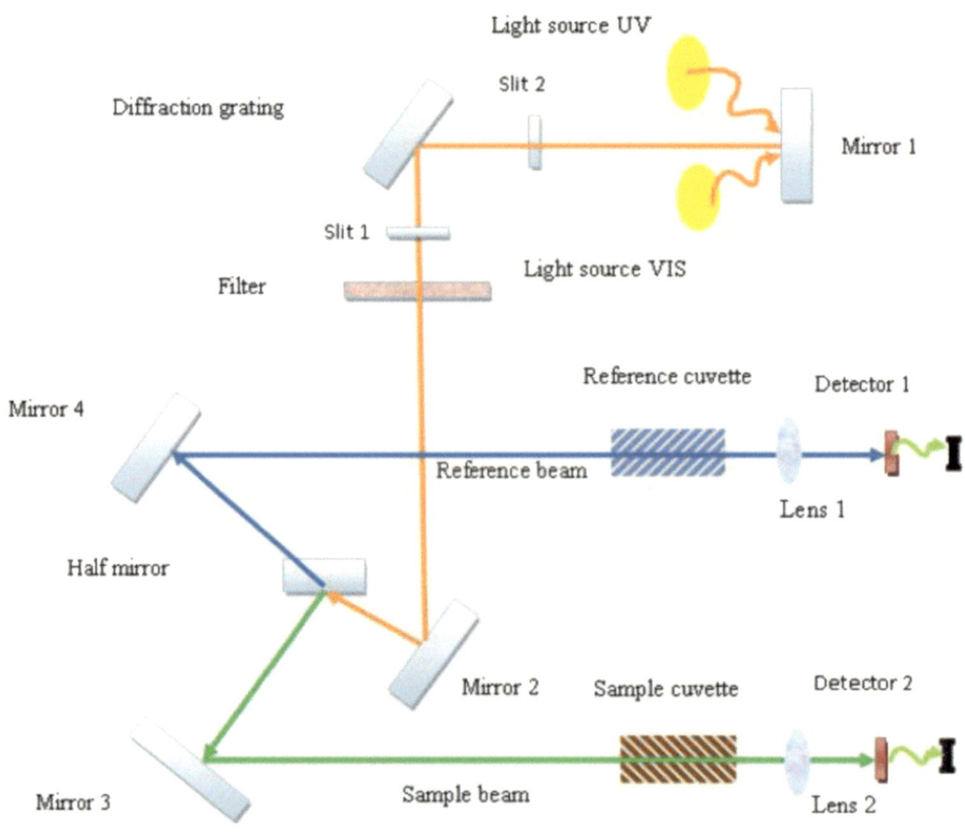

Figure 13.
Principle of operation UV–VIS spectrophotometer [11].

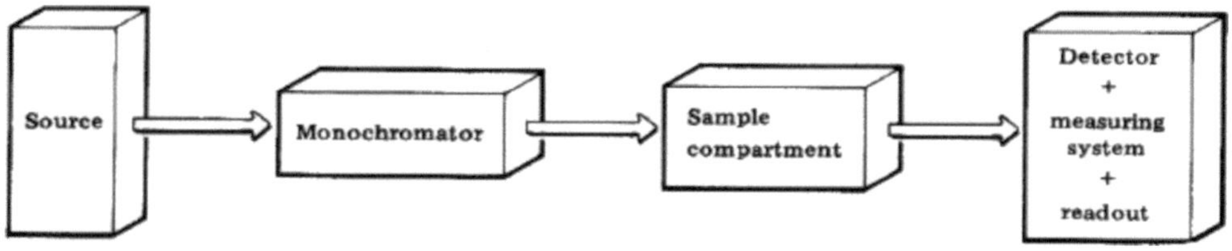

Figure 14.
Schematic drawing of spectrophotometer basic construction [12].

The minimum requirements of an instrument to study absorption spectra (a spectrophotometer) can be listed as below:

1. Radiation source with appropriate wavelengths.

2. Means of isolating light of a single wavelength and getting it to the sample compartment - monochromator and optical geometry.

3. Means of introducing the test sample into the light beam - sample handling.

4. Means of detecting and measuring the light intensity.

5.3 Spectrophotometer applications

Samples characterization by using UV/Vis spectrophotometry works by comparing the intensity light called reference (which represent the intensity of light before it passes through a sample), with the measured intensity of light passing through the sample. The ratio is called the transmittance (or absorbance), and is usually expressed as a percentage. Therefore by using a spectrometer we can measure Transmittance/Absorbance, and extract from it the following optical properties:

- Transmition/Absorption;

- Energy gap;

- Activation energy;

- Disorder;

- Refraction index.

6. Hall Effect

Hall Effect is a process in which a transverse electric field is developed in a solid material when the material carrying an electric current is placed in a magnetic field that is perpendicular to the current. Hall Effect was discovered by Edwin Herbert Hall in 1879.

Hall effect is used to determine if a substance is a semiconductor or an insulator. The nature of the charge carriers, resistivity and carrier's mobility can be measured.

6.1 Principle and theory of Hall effect

The principle of Hall Effect states that when a current-carrying conductor or a semiconductor is introduced to a perpendicular magnetic field, a voltage can be

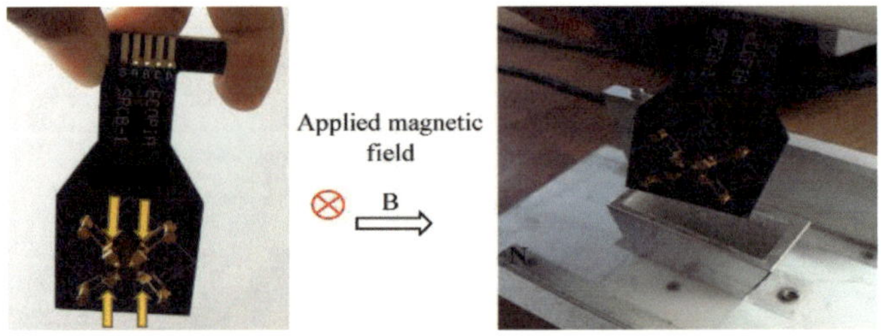

Figure 15.
Hall measurement HMS set up of CuO grown onto glass substrate [13].

measured at the right angle to the current path. This effect of obtaining a measurable voltage is known as the Hall Effect (**Figure 15**).

When a conductive plate is connected to a circuit with a battery, then a current starts flowing. The charge carriers will follow a linear path from one end of the plate to the other end. The motion of charge carriers results in the production of magnetic fields. When a magnet is placed near the plate, the magnetic field of the charge carriers is distorted. This upsets the straight flow of the charge carriers. The force which upsets the direction of flow of charge carriers is known as Lorentz force.

Due to the distortion in the magnetic field of the charge carriers, the negatively charged electrons will be deflected to one side of the plate and positively charged holes to the other side. A potential difference, known as the Hall voltage will be generated between both sides of the plate which can be measured using a meter.

The Hall voltage represented as V_H is given by the formula:

$$V_H = (IB)/(qnd) \tag{3}$$

Here,
I: is the current flowing through the sensor;
B: is the magnetic Field Strength;
q: is the charge;
n: is the number of charge carriers per unit volume;
d: is the thickness of the sensor.

6.2 Hall coefficient

The Hall Coefficient R_H is mathematically expressed as.

$$R_H = E_y/(JB) \tag{4}$$

where: J is the current density of the carrier electron, E_y is the induced electric field and B is the magnetic strength. The hall coefficient is positive if the number of positive charges is more than the negative charges. Similarly, it is negative when electrons are more than holes.

7. Binary semiconductors layers' characterization

7.1 Tin monosulfide SnS

Tin monosulfide, SnS, seems to be one of the most important and the most studied binary semiconductor compounds [14]. Indeed, it can be used in many technological devices [15]. It is specifically used in solar cells [15], as an absorber [16]. Stannous sulfide (SnS), is a very important candidate in photovoltaic application due to his high characteristics such as: direct band gap with an almost optimum value varying in the range 1.3–1.7 eV [17], high absorption coefficient ($\approx 10^5$ cm^{-1}) [18] and the orthorhombic structure [15]. In addition, it has an orthorhombic structure [15].

By using a chemical ultrasonic spray deposition method (CUS) [13, 19–21], we have elaborated two samples of SnS thin films on glass substrates. Solution of (0.07 M) molarity was prepared by mixing tin chloride $SnCl_2$ and thiourea (SC$(NH_2)_2$) as sources of Sn and S respectively. The precursors were dissolved in methanol. The experimental details are given in **Table 1**.

X-ray diffraction patterns of deposited SnS thin films at different deposition time are shown in **Figure 16**.

The films were found to be polycrystalline with a relatively strong (120) peak. In addition, the other weak peaks contained in the diffraction pattern indicates that the prepared SnS films have the orthorhombic crystal structure according to the PDF Card No. 033–1375. Other additional peaks appearing at 28.73°, 32,72° and

Deposition time (min.)	Volume (ml)	Substrate temperature (°C)	Nozzle–substrate distance (cm)
25	80	350	4.5
30			

Table 1.
Experimental conditions used for the preparation of SnS thin films.

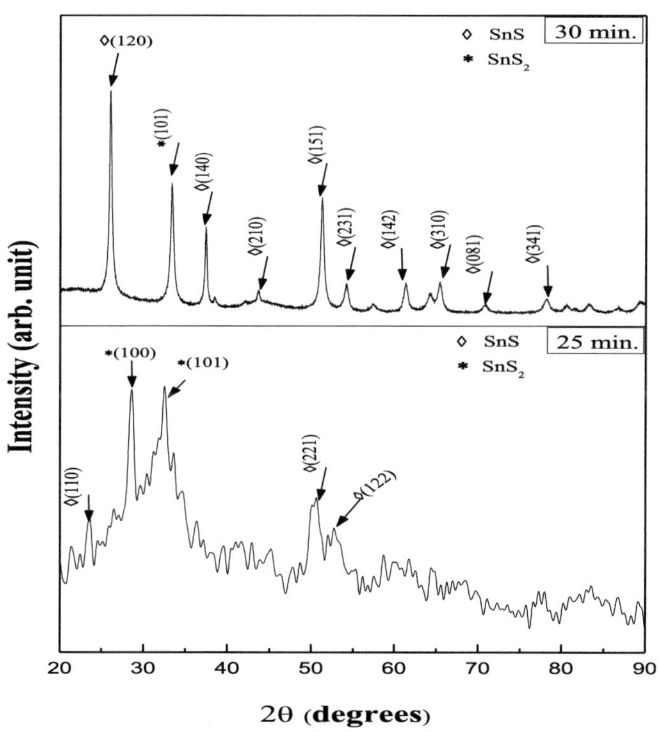

Figure 16.
XRD patterns of SnS layers deposited at different deposition time [21].

49,78° can be attributed to the SnS_2 phase (PDF Card No. 031–1399) which was found to accompany the preparation of SnS [22].

The grain sizes and internal strains of the prepared films were calculated from the XRD data using the standard Debye–Scherer formula, and are summarized in **Table 2**.

The grain size calculated from the XRD diffraction spectrum varied from 16.54 nm to 19.07 nm when the deposition time was increased from 25 min. to 30 min. Similar values of the grain size have been found by other investigators [23, 24]. These sizes values indicate the nanocrystalline nature of the films.

Films morphology and chemical composition was carried out using scanning electron microscopy (SEM) and EDX technique. The typical SEM image of SnS thin film deposited during 30 min. is shown in **Figure 16**. The film is homogeneous, devoid of cracks and has a near stoichiometric ratio (Sn/S = 1.1). This smooth aspect of the obtained film (which is consistent with the XRD) can be related to the viscosity and the surface tension of methanol. Indeed, when using methanol as solvent, the droplets are more easily spread on the substrate surface [21]. Furthermore, we notice, the presence of Al, Cl, Si and O elements which are not expected to be in films and may originate from the glass substrates (**Figure 17**).

The optical band gap is calculated using the relation

$$(\alpha h\nu)^2 = A(h\nu - E_g) \tag{5}$$

where A is a constant, Eg is the optical band gap, ν is the frequency of the incident photon and h is the Planck's constant. The plots of $(\alpha h\nu)^2$ versus the photon energy $h\nu$ for direct transition for the film deposited during 30 min. is shown in **Figure 18**. The band gap energy of this film is determined using the intercept of the tangent to the plot with the abscissa axis.

The obtained optical band gap values of the two SnS thin films are summarized in **Table 3**. We notice that the optical band gap of the prepared films decreases when the deposition time increases. This decrease of the Eg can be due the grain size increase [25].

Deposition time (min.)	D (nm)	Internal strain ($\varepsilon \times 10^{-3}$)
25	16,54	2,09
30	19,07	1,82

Table 2.
Grain sizes and internal strain of SnS films.

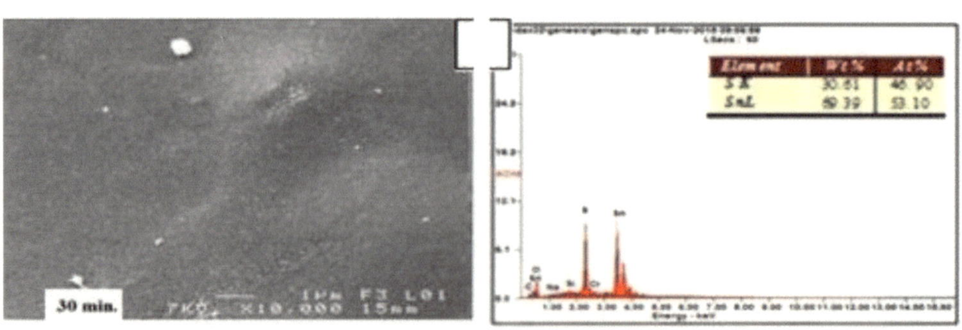

Figure 17.
SEM images and EDX spectrum of SnS film prepared during 30 min [21].

DOI: http://dx.doi.org/10.5772/intechopen.100513

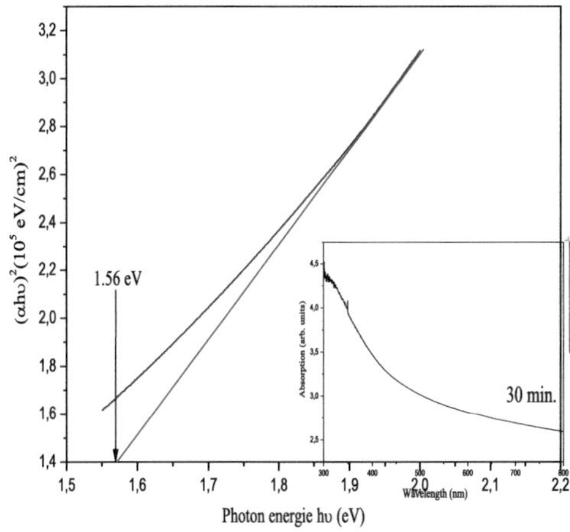

(a)

Deposition time (min.)	Optical band gap (eV)
25	1,66
30	1,56

(b)

Figure 18.
(A) Plot of $(\alpha h\nu)^2$ vs. $h\nu$ of SnS film. Inset: representation of the absorption coefficient α as a function of the wavelength. (B) Optical band gap of SnS films [21].

Deposition time (min.)	Sn / S (at.%)	Conduction type	Restivity (Ω.cm)
25	46.53 / 53.47	p-type	120
30	53.10 / 46.90	p-type	8.06

Table 3.
Electrical conductivity of of SnS films **A** *and* **B** *and their deposition time.*

In **Table 3** we report the electrical properties of the as-deposited SnS thin films.

Stannous sulfide is generally known to be a p type semiconductor [26]. Ionized tin vacancies leads to create the acceptor levels [27].

The atomic ratio obtained from the EDX analysis, conduction type and resistivity are indicated in **Table 3**. We note that SnS films have an p-type conductivity, with quasi-stoichiometric composition. Since the conductivity type of the SnS films are essentially controlled by the excess in Sn concentration in the compound [21]. In this case, the Sn atoms act as donor and lead to the n-type conduction. But due to the quasi-stoichiometric nature of our films, they exhibit a p-type conductivity. Furthermore, we note that the resistivity of the films decreases when the deposition time increases. This variation can be attribute to the grain size increase as indicated in **Table 2**. These results are consistent with the literature [18].

7.2 Tin disulfide SnS_2

Tin disulfide (SnS_2) was considered as one of very interesting tin sulfides semiconductors. SnS_2 has been known for its potential applications in solar cells as well as electrical switchings [28]. This material belongs to IV–VI group of semiconductor compound with hexagonal crystal structure (a = 0.3648 nm, c = 0.5899 nm) [29]. It has a wide band gap energy (2.88 eV) [30], and n-type electrical conductivity with magnitude depending on the preparation methods.

SnS_2 thin films were deposited onto ordinary glass substrates using the same method CUS as for SnS. The precursors used as sources of tin (Sn) and sulfur (S) are: ($SnCl_4$: $2H_2O$) and ($SC(NH_2)_2$) respectively. Two different molarities of $SnCl_4$ (M_{Sn}) were diluted with a fixed molarity (0.1 mol/l) of thiourea (M_S) in methanol, in order to study their effect on SnS_2 properties. These experimental conditions are summarized in **Table 4**.

Figure 19 shows the X-ray diffraction patterns of the elaborated films. From the obtained spectra it can be noticed, that the films are polycrystalline with a preferred orientation along the (001) direction and fit well with hexagonal SnS_2 structure according to the ASTM card number 23–0677. This indicates the presence of the pure hexagonal β-SnS_2 phase [31]. In addition, it can be noticed that, the main peak intensity increases with increasing M_{Sn} molarity.

Structural properties of the two samples are given in **Table 5**.

It can be noticed that the grain size decreases when the molarity increases, contrary to the strain. This comportment is due to the fact that grains growth is

Reagents molarity (mol/l)		Volume (ml)	Substrate temperature (°C)	Deposition time (min)
$SnCl_4$: $2H_2O$	Thiourea			
0.06	0.1	30	350	10
0.07				

Table 4.
Experimental conditions used for tin disulfide thin films elaborations.

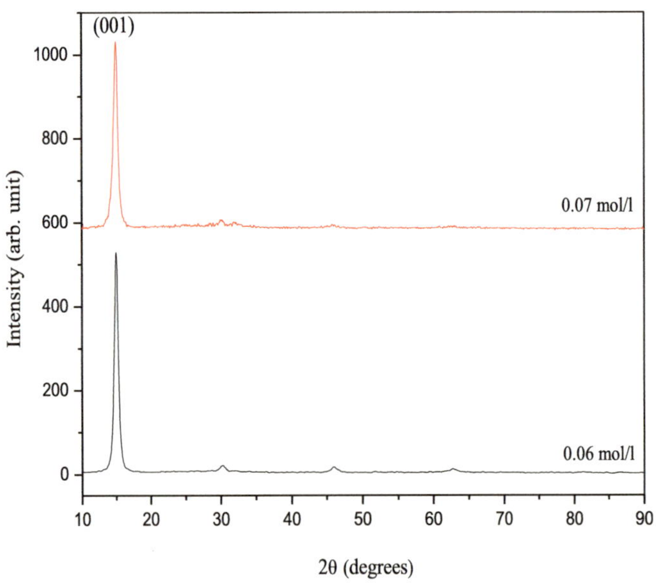

Figure 19.
XRD patterns of SnS_2 thin films deposited at different molarities [19].

Sample molarity (mol/l)	Cristallite size (nm)	Strain (*10^{-3}) %	Lattice parameters (Å)	Lattice parameters (Å) according to JCPDS card number 23–0677
0.06	16.98	2,049	a = 3.613 c = 5.889	Hexagonal structure a = 3.648 c = 5.899
0.07	12.35	2,825	a = 3.603 c = 5.873	

Table 5.
Structural parameters of dominate phase of the prepared films.

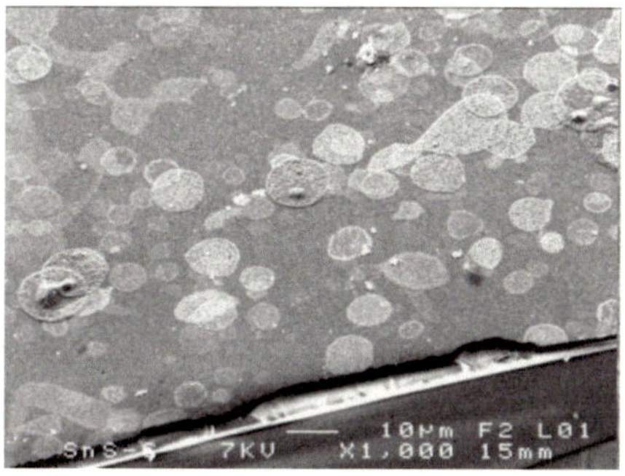

Figure 20.
SEM image of as-synthesized SnS_2 thin films [19].

controlled by the strain in film network. Because the presence of internal strain in the film network cause a minimization in the grain growth driving forces, which prevent the grain size enlargement during the film formation and vice-versa.

Morphological study of typical SnS_2 thin film (Msn = 0.07 mol/l), shows that the surface topography is dense and rough with an arbitrary distribution of the bubbles (**Figure 20**).

On the other hand, the measurement of films resistivity revealed that this latter decreases from 0.46_x10^3 Ω.cm when the molarity increases and reaches its minimum value of $0.18\ _x10^3$ Ω.cm for M_{Sn} = 0.07 mol/l.

According to the obtained results, for the two studied materials, with the different analysis techniques we can conclude that the good quality of the deposited films and the low fabrication cost of the used method can lead to solar cells whose cost-quality ratio is better than the solar cells which are fabricated by the other standard process.

Author details

Kenza Kamli* and Zakaria Hadef
Department of Physics, Faculty of Sciences, University 20 Août 1955-Skikda, Skikda, Algeria

*Address all correspondence to: kenza_kamli@yahoo.fr; k.kamli@univ-skikda.dz

References

[1] Transmission electron microscopy and diffractometry of materials "Chapter 1 Diffraction and the X-Ray Powder Diffractometer". Fultz B., Howe J.M., 2013, XX, 764 p., Hardcover. ISBN: 978-3-642-29760-1

[2] James R. Connolly, for EPS400–002, Introduction to X-Ray Powder Diffraction, Spring 2007

[3] Serna F, Lagneau J, Carpentier JM. La diffraction des rayons X : une technique puissante pour résoudre certains problèmes industriels et technologiques. Chimie Nouvelle. 2014; 116.

[4] M. Birkholz, Thin Film Analysis by X-Ray Scattering. Copyright c 2006 WILEY-VCH Verlag GmbH & Co. KGaA, WeinheimISBN: 3-527-31052-5.

[5] Sharma R, Bisen DP, Shukla U, Sharma BG. X-ray diffraction: a powerful method of characterizing nanomaterials. Recent Research in Science and Technology. 2012; 4(8): 77-79.

[6] http://www.chem.ox.ac.uk/icl/heyes/structure_of_solids/Lecture1/Bravais.gif

[7] Agarwal B.K. X-ray Spectroscopy, 2nd edn, Springer-verlag, Berlin, 1991.

[8] Goldstein, J. I., et al. Scanning Electron Microscopy and X-ray Micronalysis, 3rd ed, Plenum Press, New York, 2003.

[9] CantoniM., Introduction to EDX, MSE-636 Spring, 2014.

[10] Reimer L., Scanning Electron Microscopy, Springer-Verlag, Berlin, 1985.

[11] Kamli K., Elaboration et caractérisations physico-chimique des couches minces de sulfure d'étain par spray ultrasonique: Effet des sources d'étain, Master thesis, Mohamed Khider-Biskra University, 2013.

[12] Basic UV/Visible Spectrophotometry.

[13] Kamli K., Hadef Z., Chouial B., Hadjoudja B., Thickness effect on electrical properties of copper oxide thin films, Surface Engineering, https://doi.org/10.1080/02670844.2018.1475052, 2018.

[14] A. Sánchez-Juárez, A. Tiburcio-Silver, A. Ortiz, 'Fabrication of SnS_2/SnS heterojunction thin film diodes by plasma-enhanced chemical vapor deposition', Thin Solid Films, 2005, 480–481, 452–456.

[15] Sekhar C. Ray, Malay K. Karanjai, Dhruba DasGupta, 'Structure and photoconductive properties of dip-deposited SnS and SnS2 thin films and their conversion to tin dioxide by annealing in air' Thin Solid Films, 1999, 350, 72-78.

[16] M.M. El-Nahass, H.M. Zeyada, M.S. Aziz, N.A. El-Ghamaz, 'Optical properties of thermally evaporated SnS thin films', Opt. Mter. 2002, 20, 159-170.

[17] Jacob A. Andrade-Arvizu, M.F. García-Sánchez, M. Courel-Piedrahita, J. Santoyo-Morales, D. Jiménez-Olarte, M. Albor-Aguilera, O. Vigil-Galán, 'Pressure induced directional transformations on close spaced vapor transport deposited SnS thin films, Materials and Design, 2016, 110, 878–887.

[18] E. Guneri, C. Ulutas, F. Kirmizigul, G. Altindemir, F. Gode, C. Gumus, 'Effect of deposition time on structural, electrical, and optical properties of SnS thin films deposited by chemical bath deposition', Applied Surface Science, 2010, 257, 1189–1195.

[19] Hadef Z., Kamli K., Attaf A., Aida M. S. and Chouial B., Effect of $SnCl_2$ and $SnCl_4$ precursors on SnS_x thin films prepared by ultrasonic spray pyrolysis, Journal of Semiconductors, 38 (2017) 6.

[20] Kamli K., Hadef Z., Chouial B., Hadjoudja B., Elaboration of SnS_2 Thin Films by Ultrasonic Spray for Solar Cell Application, Global Journal of Researches in Engineering: J General Engineering, 17 (2017) 5.

[21] Kamli K., Hadef Z., Chouial B., Hadjoudja B., Zaidi B., Chibani A., Synthesis and characterisation of tin sulphide thin films, DOI: 10.1080/02670844.2016.1271593, 2017.

[22] Li Q., Ding Y., Wu H., Liu X., Qian Y., 'Fabrication of layered nanocrystallites SnS and β-SnS_2 via a mild solution route', Mater. Res. Bull., 37 (2002) 925-932.

[23] Sajeesh T.H., Jinesh K.B., Sudha Kartha C., Vijayakumar K.P., 'Role of pH of precursor solution in taming the material properties of spray pyrolysed SnS thin films', Applied Surface Science, 258 (2012) 6870– 6875.

[24] Panda S.K., Antonakos A., Liarokapis E., Bhattacharya S., Chaudhuri S., 'Optical properties of nanocrystalline SnS_2 thin films', Materials Research Bulletin, 42 (2007) 576–583.

[25] Guneri E., Gode F., Ulutas C., Kirmizigul F., Altindemir G., Gumus C., 'Properties of p-type SnS thin films prepared by chemical bath deposition', Chalcogenide Lett., 2010, 7, 685–694.

[26] Guneri E, Gode F, Ulutas C, et al. Properties of p-type SnS thin films prepared by chemical bath deposition. Chalcogenide Lett. 2010;7:685–694.

[27] Vidal J, Lany S, d'Avezac M, et al. Band-structure, optical properties, and defect physics of the photovoltaique semiconductor SnS. Appl Phys Lett. 2012;100:032104.

[28] Bissessur R., Schipper D., Exfoliation and reconstruction of SnS_2 layers: A synthetic route for the preparation of polymer-SnS_2 nanomaterials, Materials Letters 62 (2008) 1638–1641.

[29] Chakrabarti A., Lu J., McNamara A. M., Kuta L. M., Stanley S. M., Xiao Z., Maguire J. A., Hosmane N. S., Tin (IV) sulfide: Novel nanocrystalline morphologies, Inorganica Chimica Acta 374 (2011) 627–631.

[30] Tan F., Qu S., Zeng X., Zhang C., Shi M., Wang Z., Jin L., Bi Y., Cao J., Wang Z., Hou Y., Teng F., Feng Z., Photovoltaique effect of tin disulfide with nanocrystalline/amorphous blended phases, Solid State Communications 150 (2010) 58–61.

[31] Benouis C E, Benhaliliba M, Yakuphanoglu F, et al. Physical properties of ultrasonic sprayed nanosized indium doped SnO_2 films. Synthetic Metals, 161 (2011) 1509.

Chapter 4

Kesterite $Cu_2ZnSnS_{4-x}Se_x$ Thin Film Solar Cells

Kaiwen Sun, Fangyang Liu and Xiaojing Hao

Abstract

Kesterite $Cu_2ZnSnS_{4-x}Se_x$ (CZTS) is a promising thin film photovoltaic (PV) material with low cost and nontoxic constitute as well as decent PV properties, being regarded as a PV technology that is truly compatible with terawatt deployment. The kesterite CZTS thin film solar cell has experienced impressive development since its first report in 1996 with power conversion efficiencies (PCEs) of only 0.66% to current highest value of 13.0%, while the understanding of the material, device physics, and loss mechanism is increasingly demanded. This chapter will review the development history of kesterite technology, present the basic material properties, and summarize the loss mechanism and strategies to tackle these problems to date. This chapter will help researchers have brief background knowledge of kesterite CZTS technology and understand the future direction to further propel this new technology forward.

Keywords: kesterite, $Cu_2ZnSn(S,Se)_4$, CZTS, thin film solar cells, loss mechanism

1. Introduction

Thin film photovoltaic (PV) technologies such as $Cu(In,Ga)Se_2$ (CIGS) and CdTe have already demonstrated more than 20% power conversion efficiency (PEC) [1, 2] and are at their commercial stage. However, considering the Restriction of Hazardous Substances Directive (RoHS) adopted in European Union [3] and the recent classification of critical raw materials (CRM) by the European Commission [4], emerging thin film PV technologies with RoHS-compliant and CRM-free constituents are increasingly desirable. Kesterite copper-zinc-tin-selenosulfide and related quaternary semiconductor represented by chemical formula $Cu_2ZnSnS_{4-x}Se_x$ is generally accepted as one promising option for low-cost and nontoxic thin film PV.

The family of kesterite $Cu_2ZnSnS_{4-x}Se_x$ includes pure sulfide Cu_2ZnSnS_4 (CZTS), pure selenide $Cu_2ZnSnSe_4$ (CZTSe), and selenosulfide $Cu_2ZnSn(S,Se)_4$ (CZTSSe) and other related compound semiconductors. Herein we abbreviate all $Cu_2ZnSnS_{4-x}Se_x$ compounds as CZTS. The formation of CZTS is derived from cation mutations (cross-substitutions) in $CuInS_2$ by replacing two In atoms with one Zn and one Sn [5, 6]. CZTS possess high absorption coefficient of over 10^4 cm^{-1}, tunable band gap that can range from 1.0 to 1.5 eV to favorably match the solar spectrum, intrinsic p-type conductivity, and a three-dimensional symmetry of carrier transport [7–9]. These decent photovoltaic properties of CZTS have attracted considerable attention and enabled its rapid development in the last decades. The highest power conversion efficiencies (PCEs) of Cu_2ZnSnS_4, $Cu_2ZnSnSe_4$

and $Cu_2ZnSn(S,Se)_4$ have been set to 11%, 12.5%, and 13%, respectively [10–12], which represent the best PCE among the emerging RoHS-compliant and CRM-free inorganic thin-film PV technologies.

In this chapter, the development history and current status of CZTS PV technology will be briefly introduced. The basic physical and chemical properties of CZTS thin film will be described to help readers have a better understanding of the material. The device architecture and absorber processing will be touched; finally. The limiting factors as well as the perspective for future development of this technology will also be reviewed and addressed.

2. The development history of CZTS

The CZTS single crystal was first grown by Nitsche, Sargent, and Wild in 1966 when they tried to prepare a serious of $A^I_2B^{II}C^{IV}X_4$-type quaternary chalcogenides using iodine vapor transport [13]. The photovoltaic effect of CZTS was exhibited for the first time in 1988 by Ito and Nakazawa on a heterojunction diode consisting of cadmium-tin-oxide transparent conductive layer and CZTS thin film on a stainless steel substrate. The open-circuit voltage was measured to be 165 mV under AM1.5 illumination [14]. The open-circuit voltage was increased to 250 mV after annealing the device in air, and short-circuit current of 0.1 mA/cm^2 was achieved [15].

The first CZTS solar cell with PCE of 0.66% was reported by Katagiri et al. in 1996 at PVSEC-9, with the device structure of ZnO:Al/CdS/CZTS/Mo/soda lime glass (SLG) substrate [16]. The CZTS thin film was fabricated by vapor-phase sulfurization of E–B-evaporated precursors [17]. In 1997, Friedlmeier et al. reported the CZTS solar cell with PCE of 2.3% and open-circuit voltage of 570 mV based on thermal evaporation [18]. In 1999, the Katagiri group improved the PCE up to 2.63% [19], after that they did a lot of work on optimizing the CZTS thin film and pushed the efficiency to 5.74% until 2007 [19–24].

The CZTS solar cell started to gain intensive interest from academic and industry community from 2008 when its efficiency was further boosted to 6.7% by Katagiri et al. and when the CIGS solar cell became mature in its commercialization stage [25]. A lot of research institutes and solar cell manufactures such as Toyota, IBM, NREL (National Renewable Energy Laboratory), Solar Frontier, EMPA (Swiss Federal Laboratories for Materials Science and Technology), HZB (Helmholtz-Zentrum Berlin), ZSW (Centre for Solar Energy and Hydrogen Research Baden-Württemberg), UNSW (University of New South Wales), and so on involved in the development of this technology and significant advances have been achieved in the following decade [10, 26–33]. During the development period, Se incorporation has attracted significant attention and led to impressive progress in PCE [34–37]. Moreover, various fabrication methods including vacuum deposition process and non-vacuum process such as solution method, electrochemical deposition, etc., have been developed for fabricating the CZTS thin film [29, 37–40].

An important milestone in the development of CZTS solar cell is the 10% benchmark efficiency breakthrough achieved by IBM Thomas J. Watson Research Center in 2011 [34], which shows substantial commercial promise for the CZTS-based class of thin film PV materials. This breakthrough also established the leading position of IBM in CZTS PV technology research, represented by a serious of world record efficiencies [38, 39, 41]. The highest PCE for CZTS solar cell has been stagnant at 12.6% for more than 6 years since 2014 when IBM last updated their efficiency breakthrough [39]. Thanks to the better understanding of the CZTS material and the loss mechanism of the device, quite a few groups reported CZTS PCE close to the 12.6% record efficiency in recent years [11, 42–46]. The recently announced NREL Best Research-Cell Efficiency Chart included the newly refreshed CZTS record efficiency of 13% achieved by Xin et al. from NJUPT [12]. This small step breakthrough comes from great efforts in this research area and hopefully will bring more interests and confidence to the CZTS R&D community.

3. The physical and chemical properties of CZTS thin film

The investigation and understanding of nature of the CZTS thin film are crucial for further developing this technology. Intensive studies have been conducted during the rapid development stage; therefore, the physical and chemical properties of CZTS have been well revealed.

3.1 Crystal structure

As we briefly introduced earlier, CZTS is derived from the cation mutations of $CuInS_2$ (CIS), while both are originated from the binary II–VI semiconductors adopting the cubic zinc-blende (or hexagonal wurtzite) structure as shown in **Figure 1**.

Similarly, the structure of CZTS is also derived from ternary I-III-VI_2 compounds. In general, as shown in **Figure 2**, chalcopyrite (CH) and CuAu-like (CA) structures are two fundamental I-III-VI2 structures that obey the octet rule [5, 47, 48]. Therefore, the quaternary CZTS is well known as two principal structures, kesterite (space group ***I*4**, **Figure 2d**) and stannite-type (space group ***I*42m**, **Figure 2e**), which are derived from CH structure and CA structure, respectively. Other primitive mixed CA structure (PMCA) (space group ***I*42m**, **Figure 2f**) derived from CA structure has also been reported in CZTS [5, 47, 48]. As the most common structures discussed in literature are kesterite and stannite-type, we only focus on these two structures in this section. The two structures are closely related with the main difference of cation arrangement. Both structures are composed of a cubic close-packed lattice of S anions, with half of the tetrahedral interstices occupied by cations. Sn atoms occupy the same fixed positions in both structures, but the Cu and Zn atoms are in different position [49, 50]. In kesterite structure, cation layers of CuZn, CuSn, CuZn, CuZSn alternated at z = 0, ¼, ½, and ¾ respectively (**Figure 2d**), while in stannite structure, ZnSn and Cu_2 layers alternate with each other (**Figure 2e**). The similarities of structure make it difficult to distinguish kesterite and stannite experimentally by employing common X-ray diffraction and

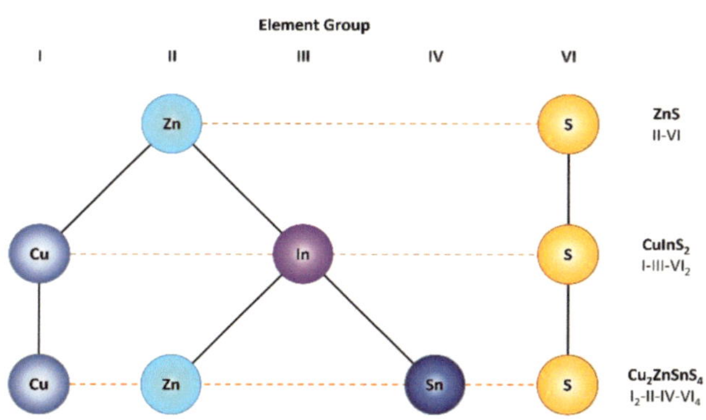

Figure 1.
Schematic illustration of the origin of CZTS structure. Reproduced from [6] with permission from the Royal Society of Chemistry.

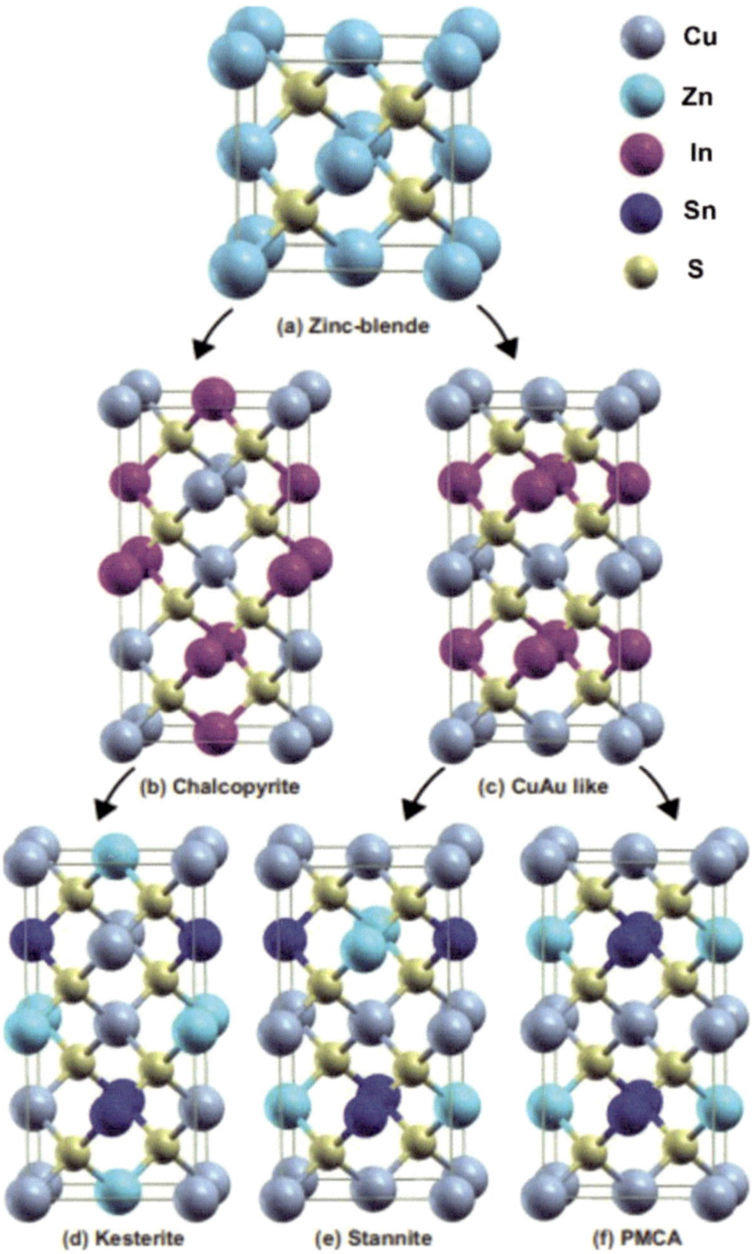

Figure 2.
The crystal structure of (a) zinc-blende ZnS, (b) chalcopyrite $CuInS_2$, (c) CuAu-like $CuInS_2$, (d) Kesterite-type Cu_2ZnSnS_4, (e) stannite-type Cu_2ZnSnS_4, and (f) PMCA-Cu_2ZnSnS_4. Reproduced from [47] with permission from the American Physical Society.

Raman spectroscopy techniques [51, 52]. Only advanced techniques such as neutron powder diffraction analysis are capable to tell them apart [50, 53].

CZTS usually exists as kesterite-type structure, which is more stable thermodynamically than the stannite-type [54]. This is in agreement with the experimental observation [21, 50, 55, 56]. Plenty of theoretical studies have also confirmed that the kesterite-type structure is the ground state structure in CZTS [5, 47, 48, 57–59]. This is also the reason that CZTS is named after Kesterite because it crystallizes as kesterite structure. However, the energy difference between the kesterite and the stannite-type structure is rather small [5, 47, 48, 57–59]. This indicates that kesterite structure should be formed under equilibrium growth conditions, but both phases may exist, especially when growth method and conditions are changed, it should be relatively easy to grow materials with mixed phases. This may also partly explain the existence of disorder structure with more random distribution of the Cu and Zn on the cation positions [60], which will be discussed in more details in the following sections.

3.2 Electronic band structure

Kesterite CZTS has a similar tetrahedral bonds geometry as traditional group I–V, III–V, and II–VI semiconductors. It obeys Lewis' octet rule with eight electrons around each anion atom (S or Se); the four bonds of each anion therefore form together a close valence shell [48]. However, there are fundamental differences between the Cu-based quaternary compound and the group I–V, III–V, and II–VI binary semiconductors. First of all, the bonds in CZTS involve Cu-d–anion-p hybridized antibonding states. This weakens the bonds in CZTS. Second, Cu in CZTS has only one valence s-like electron while the group III–V (group II–VI) semiconductors have three (two) valence s-like electrons for their cations.

Several first principle studies have revealed the electronic band structure of CZTS [47, 48, 58, 59, 61, 62]. An example is as shown in **Figure 3**, the calculated band structures of kesterite Cu_2ZnSnS_4 and $Cu_2ZnSnSe_4$ from different methods are presented. Overall, the kesterite CZTS materials are all direct-gap semiconductors. Different calculation methods generated slightly different values, for example, generalized gradient approximation (GGA) method calculated the band gap energy for Cu_2ZnSnS_4 and $Cu_2ZnSnSe_4$ as 1.56 eV and 1.05 eV, respectively. Corresponding band gap values from hybrid functional calculations (HSE06) are 1.47 and 0.90 eV, while the values from the GW_0 calculation are 1.57, and 0.72 eV. These values are in agreement with other calculated data and available experimental measurement, converging to the results that $Eg \approx 1.5$ eV in CZTS and $Eg \approx 1.0$ eV in CZTSe [14, 24, 28, 64–72] to vary linearly as a function of the Se content x [73] with Eg = 1.47, 1.30, 1.17, 1.01, and 0.90 eV for x = 0, 1/4, 1/2, 3/4, and 1, using the HSE06 potential. This linear relationship also agrees with experimental results [66, 70, 71].

The density of states (DOS) of kesterite Cu_2ZnSnS_4 and $Cu_2ZnSnSe_4$ can also be calculated as illustrated in **Figure 4** [63]. It is not surprising to see that Cu_2ZnSnS_4 and $Cu_2ZnSnSe_4$ show comparable DOS as they are in same tetrahedral bond geometries. The DOS of conduction bands in Cu_2ZnSnS_4 is ~0.5 eV higher than that of $Cu_2ZnSnSe_4$ because of large energy gap. It is common that in Cu-based chalcogenides including the quaternary (for example, CIS) and ternary compounds (for example, CZTS), the valence band maximum (VBM) is derived mainly from the hybridization of anion p and Cu d states because Cu has higher d orbital energy than Zn, Ga, In, and Sn [59, 74]. In CZTS, the valence p level of S is lower in energy than Se, thus the VBM of the sulfides is lower than that of the selenides. This difference is reduced by anion p–Cu d overlap (p-d hybridization) because the hybridization is stronger in the shorter Cu–S bond and pushes the antibonding VBM level of the sulfide up relative to that of the selenide. Therefore, the valence band offset between the Cu_2ZnSnS_4 and $Cu_2ZnSnSe_4$ is less than 0.2 eV. The DOS of the CZTS conduction band minimum (CBM) is primarily controlled by the Sn-s and anion p-like states due to the lower s orbital energy of Sn than the other cations [59, 74]. More importantly, in CZTS the lowest CB is separated from the higher energy bands and is therefore more localized in energy. Therefore, the CBM of CZTS is expected to vary depending on the alloying on the Sn-site with other group-IV elements

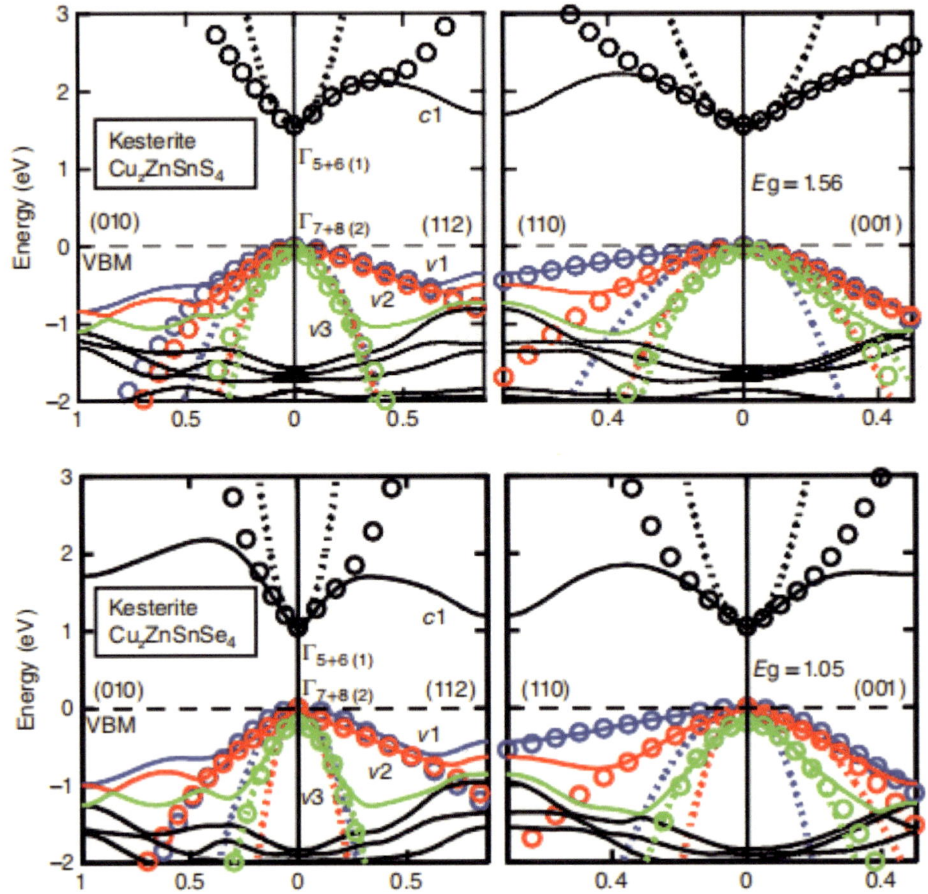

Figure 3.
The electronic band structure of the kesterite structures of Cu_2ZnSnS_4 and $Cu_2ZnSnSe_4$ along four symmetry directions. The energy refers to the VBM (dashed lines). The spin–orbit interaction is included, but the index of the bands refers to spin-independent bands where c1 represents the lowest CB and v1 represents the topmost VB. Different line shapes represent different calculation methods [63]. Reproduced from [59] with permission from American Institute of Physics.

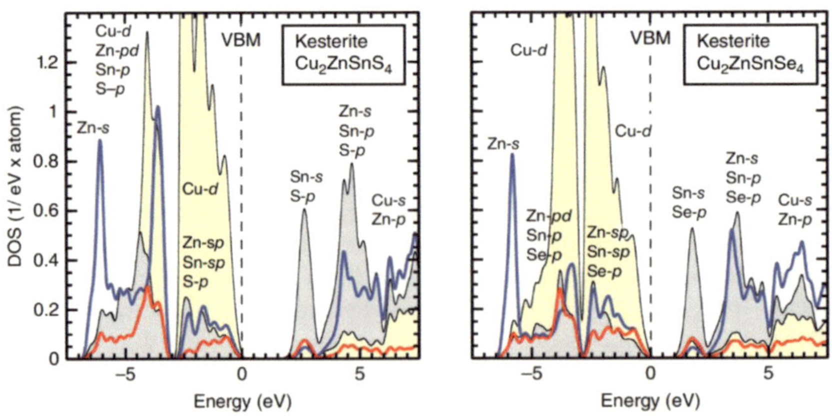

Figure 4.
Atomic and angular momentum resolved DOS of kesterite Cu_2ZnSnS_4 and $Cu_2ZnSnSe_4$ presented with a 70 meV Lorentzian broadening. Reproduced from [59] with permission from American Institute of Physics.

(e.g., Ge) due to this localized CB. The band gap energy can therefore be tailored by cation alloying for an optimized optical efficiency of the materials.

3.3 Composition variation and phase competition

As a quaternary semiconductor, CZTS consists of three metals and up to two chalcogens, which provides a wide range for composition variation and secondary phase formation. However, high-efficiency CZTS thin film solar cells require single-phase kesterite absorber, and some secondary phases have been reported detrimental to the device performance. This section reviews the stable phase region of CZTS and possible secondary phases likely to be formed in the quaternary system.

The 3D CZTS quaternary phase diagram as shown in **Figure 5** can be deduced from Cu_2S-ZnS-SnS_2 pseudo-ternary diagram [77] and CuS-ZnS-SnS pseudo-ternary phase diagram [75] (selenide kesterite system also refers to Cu_2SnSe_3–$SnSe_2$–ZnSe diagram [78]. In addition, the stability region of the CZTS in the atomic chemical potential space can be calculated as shown in **Figure 5** [76, 79]. All these diagrams indicate that volume of the stable CZTS region is small, and the slight deviation (maximum 1–2 at%) outside this space will cause the formation of different secondary phases [5, 80]. The narrow stable window also implies that the composition control and the chemical potential control are crucial for the growth of high-quality and single-phase CZTS absorber. The Zn content is particularly important because the stable region along the μ_{Zn} axis is much narrower. Moreover, it is well accepted experimentally and theoretically that high-efficiency solar cells need CZTS absorber with Zn-rich and Cu-poor composition [8–10, 27, 28, 33, 67, 81]. This makes it more difficult to control the secondary phases formation.

According to the phase diagram and literature report, the most common secondary phases in the Cu-Zn-Sn-S/Se system are list in **Table 1** [92]. The influence of the secondary phases on the solar cell performance depends on their position in the film as well as their physical properties. For example, Cu_2S(Se) phases in the final film may act as shunting path due to both the high conductivity and contact with front and back interfaces. However, Cu_2S(Se) is also an important fluxing agent to promote lateral grain growth during the film growth [28]. Generally, If the secondary phase has a lower band gap than the CZTS absorber, it will limit the open-circuit

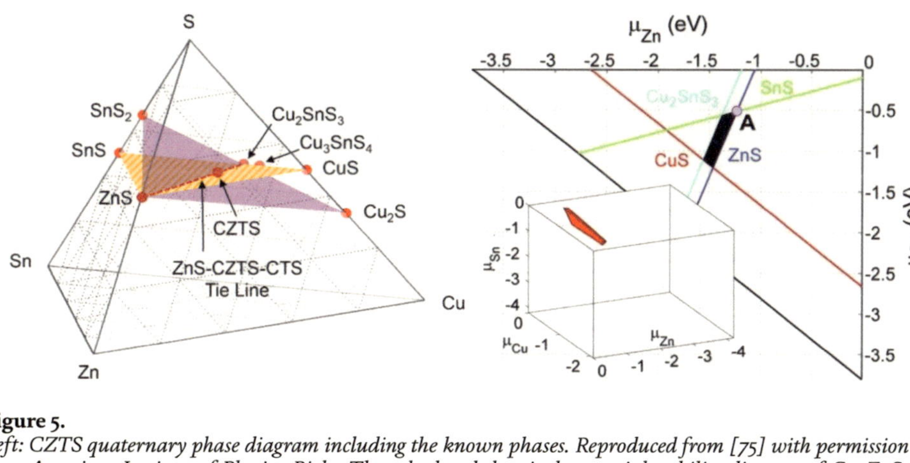

Figure 5.
Left: CZTS quaternary phase diagram including the known phases. Reproduced from [75] with permission from American Institute of Physics. Right: The calculated chemical-potential stability diagram of Cu2ZnSnS4 in a 2D Cu-rich plane (the stable 3D region is inset) reproduced from [76] with permission from American Institute of Physics.

Compound	Band gap (eV)	Ref.	Compound	Band gap (eV)	Ref.
Cu_2ZnSnS_4	1.5	[80]	$Cu_2ZnSnSe_4$	1.0	[80]
Cu_2SnS_3	1.0	[82]	Cu_2SnSe_3	0.8	[83]
ZnS	3.7	[84]	ZnSe	2.7	[84]
SnS_2	~2.5	[85]	$SnSe_2$	1.0–1.6	[86]
SnS	1.0 indirect, 1.3 direct	[87, 88]	SnSe	1.3	[89]
Cu_2S	1.2	[90]	Cu_2Se	1.2	[91]

Table 1.
Most common secondary phases in the Cu-Zn-Sn-S/Se system.

voltage of the solar cell. While secondary phases with higher band gaps than CZTS are less detrimental; however, they can block the transport when present in large amounts [93] or at least increase the series resistance [94]. ZnS(e) is the most likely secondary phase considering the phase diagram and especially when adapting the Zn-rich and Cu-poor condition. Fortunately, ZnS(e) is a large band gap compound and is expected to be rather benign if present in small amounts. It has even been reported that ZnS with similar crystalline structures to CZTS may passivate grain boundaries or heterojunction interface [81, 95, 96] by reducing strain and lowering recombination velocities at the grain interfaces. The tin compounds are unlikely to form because they are usually volatile [97] and will evaporate in most preparation conditions. As shown in **Table 1**, one unfavorable low band gap secondary phase in both Cu_2ZnSnS_4 and $Cu_2ZnSnSe_4$ system is the ternary $Cu_2SnS(e)_3$, which is also one reason for optimal composition range (Zn-rich and Cu-poor) found for the best solar cells. However, the non-homogeneous composition across the CZTS film under normal preparation conditions still provides the possibility to form such detrimental secondary phases [98]. Therefore, how to control the amount and position of the secondary phases in the CZTS absorber needs to be elaborately studied.

3.4 Lattice defects

The formation and properties of lattice defects are important parameters of semiconductor materials and are crucial to the function of photovoltaic devices, because they directly influence the generation, separation, and recombination of electron–hole pairs. The lattice defects (e.g., vacancies, interstitials, antisites) in kesterite CZTS system are complicated because of the increased number of component elements and the similar cation size as well as small chemical mismatch of Cu^+ and Zn^{2+}. In this section, the formation and ionization of the lattice defects in kesterite will be briefly reviewed. More importantly, the underlying mechanism of p-type conductivity, Zn-rich and Cu-poor condition growth condition, as well as limiting factors for device performance in kesterite CZTS solar cells from the perspective of lattice defects will be discussed.

The concentration of the lattice defects is determined by their formation energy. In **Figure 6**, the calculated formation energies of different defects are plotted as functions of the Fermi energy (0 means that the Fermi energy is at VBM, while 1.5 or 1.0 eV means that Fermi energy is at CBM). It is obvious that in both Cu_2ZnSnS_4 and $Cu_2ZnSnSe_4$, Cu_{Zn} antisite is the lowest energy defects, which is different from the defect properties of their parent compounds ($CuInSe_2$ or $CuGaSe_2$) where the dominant defect is the Cu vacancy V_{Cu} [76, 79, 99]. In addition, the formation energies of most acceptor defects are lower than those of donor defects, explaining the intrinsic p-type conductivity observed in literature [21, 48, 64, 84, 100–110].

DOI: http://dx.doi.org/10.5772/intechopen.101744

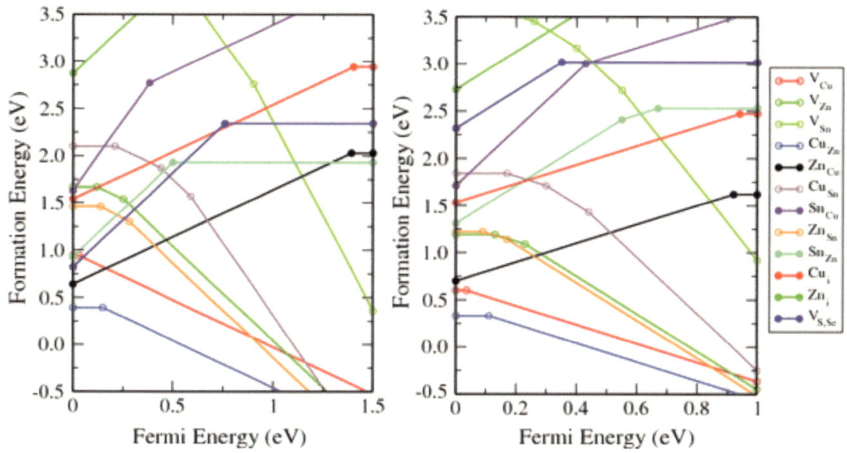

Figure 6.
Calculated defect formation energy as a function of the Fermi energy at the thermodynamic chemical-potential point a (from ***Figure*** *5 right) for Cu_2SnZnS_4 (left) and $Cu_2SnZnSe_4$ (right). For each value of the Fermi energy, only the most stable charge state is plotted, with the filled circles (change of slope) representing a change in charge state (transition energy level). Reproduced from [76, 79] with permission from American Institute of Physics and American Physical Society.*

Another important parameter of lattice defects is their ionization (transition) levels, which determines if they can produce free carriers and contribute to the electrical conductivity. The calculated ionization levels of intrinsic defects in the band gap of Cu_2ZnSnS_4 and $Cu_2ZnSnSe_4$ are shown in **Figure 7**. First of all, in both cases, the dominant defect Cu_{Zn} has an acceptor level (0.11 eV and 0.15 eV above VBM in $Cu_2ZnSnSe_4$ and in Cu_2ZnSnS_4, respectively) deeper than that of V_{Cu}. The relatively deep level of dominant antisite defect in CZTS is not favorable to the device performance because it will limit the open-circuit voltage. This also partially explains why Zn-rich and Cu-poor condition has normally been found to beneficial to solar cell efficiency because it could decrease the formation energy and enhance the population of shallow V_{Cu}. The other acceptor defects (e.g., Cu_{Sn}, Zn_{Sn}, V_{Zn}, and V_{Sn}) have higher formation energy; therefore, they have negligible contribution to the p-type conductivity. However, they may act as recombination centers especially for those deep transition levels such as (4−/3-) and (3−/2-) in band gap as seen in **Figure 7**.

In addition to the point defects, various self-compensated defect clusters can be formed in CZTS due to the large amount of low-energy intrinsic defects. Defect compensation in ternary CIS is well known to have electrically benign character because the intrinsic defects undergo self-passivation through the formation of defect complexes such as $[2V_{Cu}^{-} + In_{Cu}^{2+}]$. Therefore, it is also interesting to see if the same behavior can be observed in quaternary kesterites. Cu_{Zn} and Zn_{Cu} are the lowest-energy acceptor and donor defects, respectively; therefore, antisite pair $[Cu_{Zn}^{-} + Zn_{Cu}^{+}]$ also shows extremely low formation energy. Fortunately, its impact on the electronic structure and optical properties is relatively weak. The detrimental defect clusters are those composed of deep level defects, such as Sn_{Zn}, Sn_{Cu}, Cu_{Sn}, and Zn_i. For example, clusters with Sn_{Zn} induce large conduction band edge downshift, which could limit the solar cell performance, because the induced states are deep and may trap photo-generated electrons from the high conduction band. $[2Cu_{Zn} + Sn_{Zn}]$ clusters could present in high population if in single-phase CZTS (Cu/(Zn + Sn) and Zn/Sn ratios near 1) chemical potential conditions and be detrimental to the solar cell performance. The Zn-rich and Cu-poor condition could prevent the formation of $[2Cu_{Zn} + Sn_{Zn}]$ clusters because its formation energy

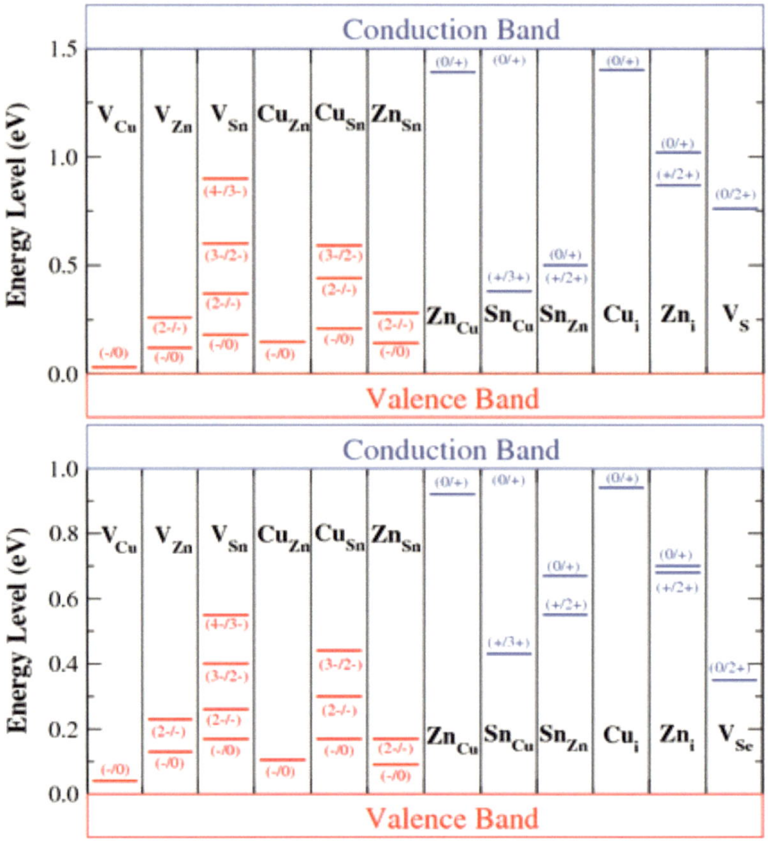

Figure 7.
The ionization levels of intrinsic defects in the band gaps of Cu_2ZnSnS_4 (top) and $Cu_2ZnSnSe_4$ (bottom). The red bars show the acceptor levels, and the blue bars show the donor levels, with the initial and final charge states labeled in parentheses. Reproduced from [79] with permission from American Physical Society.

is very sensitive to the chemical potential of Zn. This again partly explains that the Zn-rich and Cu-poor condition is beneficial to solar cell efficiencies from the defect perspective.

4. The path towards high-efficiency kesterite solar cell

4.1 Device architecture

The typical device architecture (**Figure 8**) of kesterite-based solar cell inherits from its predecessor chalcopyrite $CuInSe_2$ solar cells due to their similar optical and electronic properties. Metal Mo layer with thickness of 500 nm–1000 nm is usually deposited with sputtering on soda lime glass or other substrates as back contact. Then the kesterite CZTS absorber layer is deposited on top of the Mo layer. The p-n junction is formed by p-type CZTS and the following deposited n-type buffer layer. Typical n-type buffer layer in kesterite and chalcopyrite solar cell is 50 nm–100 nm thick CdS layer usually by chemical bath deposition. Alternative buffer materials such as ZnSnO, Zn(O, S), ZnCdS, etc., have been explored, especially in pure sulfide Cu_2ZnSnS_4 to tackle the unfavorable band alignment found at p-n junction. Next, a 50 nm–100 nm high-resistive intrinsic ZnO (i-ZnO) is deposited.

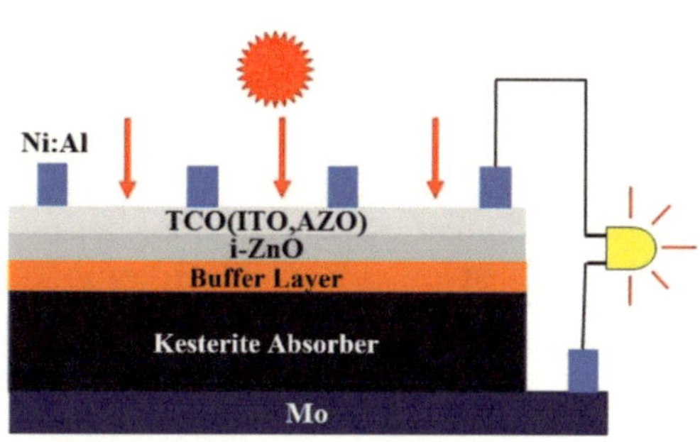

Figure 8.
Schematic device structure of a typical kesterite Cu_2ZnSnS_4 thin film solar cell.

Subsequently, the device structure is completed by deposition of 200 nm–500 nm thick Al-doped ZnO (AZO) or indium tin oxide (ITO) transparent conducting oxide (TCO) layer as the front contact. Ni/Al metal contacts are deposited on the TCO layer for improved current collection. Antireflection coating such as MgF_2 is often deposited on top of the cell to reduce the reflection loss.

Various deposition methods have been investigated for producing high-quality CZTS layer during the development of this technology. These deposition methods are broadly classified as vacuum-based and non-vacuum-based techniques. The vacuum-based methods are usually considered easy to be expanded to commercial scale because of the precise process control. All physical vapor deposition techniques including thermal evaporation, E-beam evaporation [17], sputtering [10, 33, 81, 111], pulsed laser deposition [112] fall into this category. The non-vacuum-based methods are always regarded as low-cost, high-throughput techniques and feasible in roll-to-roll production. These methods usually involve chemical or solution process, such as electrochemical deposition [40], nanoparticle-based synthesis [113, 114], sol–gel spin coating [44, 115], chemical bath deposition (CBD) [116], successive ion layer adsorption and reaction (SILAR) [117], screen printing [118], etc.

4.2 Loss mechanism and solutions

4.2.1 Open-circuit voltage (V_{OC})

In kesterite CZTS solar cells, it has been widely accepted that V_{OC} loss accounts for the majority (more than 50% [119]) of the efficiency gap between the best performance device and the theoretical limit (i.e., Shockley–Queisser radiative limit [120]). V_{OC} loss is determined by the recombination path in the device. The dominant recombination can occur in the bulk of the absorber in the quasi-neutral zone as band-to-band recombination, via defects, or in the space charge region. Another important recombination path can be located at the heterojunction interface between the buffer/window layer and absorber.

Therefore, the absorber problems discussed in **Section 3** such as abundant point defects and defect clusters (i.e., cation disordering), composition variation with narrow stable region are believed to contribute to the recombination. The Cu and Zn substitution with low energy causes a large population of antisite defects such as Cu_{Zn} and Zn_{Cu} and related defects complexes. Consequently, severe electrostatic potential fluctuation and associated band tailing can be observed [121]. In addition, the microinhomogeneities in composition, nonuniform strain, as well as formation of secondary phases lead to band gap fluctuations, which are also detrimental to

V_{OC}. Moreover, acceptor-like Cu_{Zn} defects at the interface will cause Fermi level pinning to a low energy level [122], thus reducing the band bending in the absorber and weakening the electric field. To alleviate the recombination occurring in absorber layer, great efforts have been made over recent years. One solution that has been intensively investigated is cation substitutions because the introduction of larger size cations could result in better cationic ordering, thereby reducing the defects formed because of the Cu and Zn substitution. Using Ag as a substitute for Cu [115], Cd as a substitute for Zn [123, 124], and Ge substitute for Sn [125] are popular choices as they are picked from the same cation groups. As shown in **Figure 9**, Promising progress has been made in this direction both experimentally and theoretically, demonstrating several remarkable efficiencies and mechanism of the defect emerging with cation substitution. Another possible solution to tackle the cationic disorder and related band tailing, as well as to activate the shallower defect is deliberate control of the synthesis condition, for example, post annealing the CZTS absorber within the critical temperature range of 200–250°C and reasonable time range of 1–4 hours will improve the CZTS lattice ordering and increase the band gap [126, 127]. Control of the precursor fabrication process such as the metal stack order, the valence states of the Sn source, has also been reported effective in obtaining less defective absorber and corresponding high-efficiency device [128, 129]. Modification of the sulfurization or selenization condition could also suppress the formation of detrimental intrinsic defects and defect clusters by creating a desirable local chemical environment [11].

In term of the interface recombination path, it is mainly caused by the unpassivated charge defects at the heterojunction and/or the undesirable band alignment between the CZTS absorber and conventional CdS buffer layer. The band alignment problem is more prominent in high band gap pure sulfide Cu_2ZnSnS_4 because of its higher conduction band position. A serious of alternative wide band gap buffer layer materials including ZnCdS [33], Zn(O,S) [130, 131] and ZnSnO [132, 133] have been reported effective in mitigating the band alignment issues. The charge defects at the interface can be reduced by introducing ultrathin layer that has better lattice match with CZTS [10, 81] or alloying Ag near the heterojunction interface to from intrinsic or weak n-type $(Cu,Ag)_2ZnSn(S,Se)_4$ [115, 134]. Dielectric layers have also been investigated for passivating the interface defects, thereby suppressing the interface recombination [111, 135].

4.2.2 Short-circuit current density (J_{SC})

While the V_{OC} loss contributes dominant performance loss in kestrite solar cells, J_{SC} also represents a significant limiting factor for efficiency increases. The J_{SC} loss is mainly caused by two reasons: one is the light or photons that is reflected or absorbed by layers above the CZTS (such as the buffer or window layers as shown in **Figure 8**), which cannot generate electron–hole pairs. The other is the low carrier collection efficiency due to the short diffusion length or narrow depletion width. The first problem can be addressed by introducing antireflection coating and optimizing the optical designs of the top layers above CZTS to increase the fraction of incident light reaching CZTS. The solution has been proven feasible when applying wide band gap buffer layer [111, 133], reducing the thickness and roughness of the top layers [119]. The typical External Quantum Efficiency (EQE) curve of CZTS usually shows relative lack of long wavelength response [9], which might be related to the low carrier lifetime. This low lifetime could also be the results of high defect density in the absorber layer or could simply be a consequence of high recombination loss at the back contact or at the front interface. Therefore, the solutions for suppressing the formation of detrimental defects disused in V_{OC} loss above are also

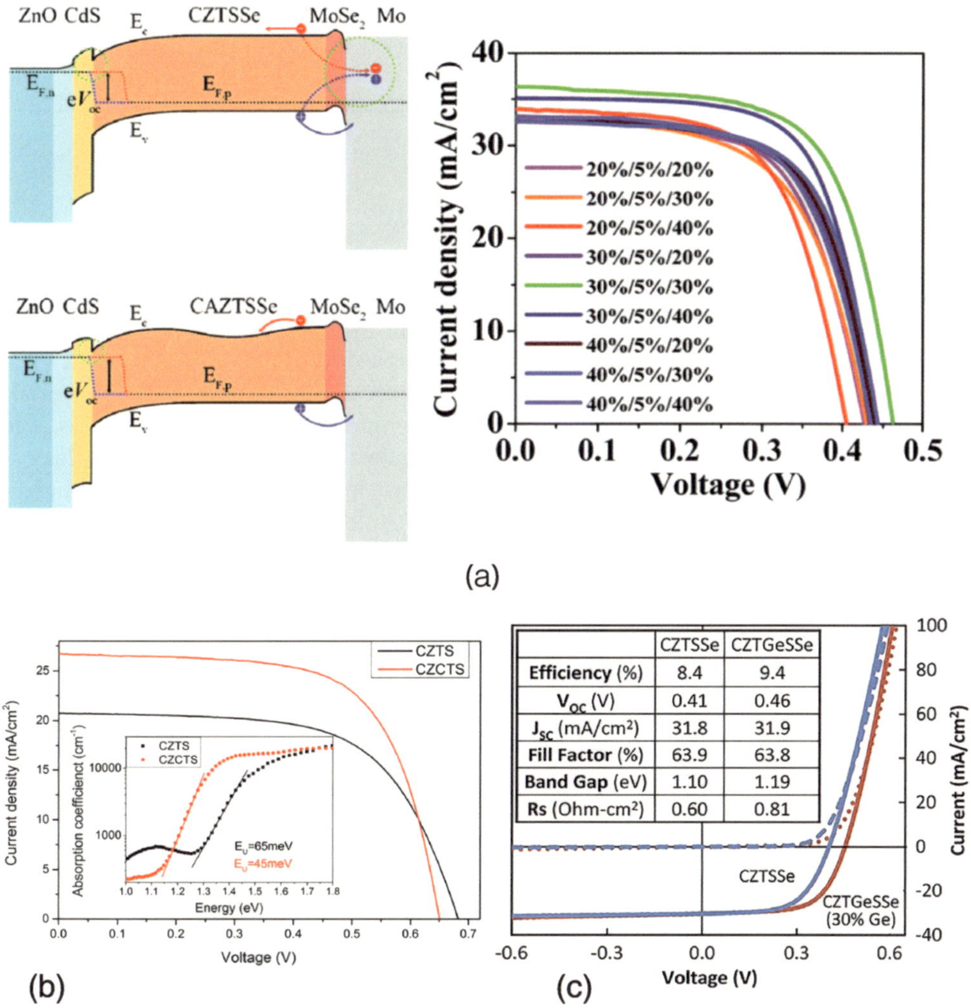

Figure 9.
(a) The band diagrams of CZTSSe and Ag-graded $(Cu_{1-x}Ag_x)_2ZnSn(S,Se)_4$ (CAZTSSe) solar cells under AM 1.5G 100 mW cm2 illumination and the J–V curves of CAZTSSe solar cells with different Ag composition gradients from the CdS interface to the Mo substrate. (b) J–V curves of CZTS and champion $Cu_2Zn_xCd_{1-x}SnS_4$ (CZCTS) devices and absorption coefficients and Urbach energy (EU) obtained from PDS measurement. (c) J–V curves for the champion CZTSSe and CZTGeSSe (30% Ge) solar cells. Reproduced from [115, 124, 125] with permission from Royal Society of Chemistry, American Chemical Society and John Wiley and Sons.

good for the J_{SC} improvement In addition, band gap grading has also been developed in CZTS to increase the EQE by controlling the [S]/[S + Se] ratio at the surface and back surface [136].

4.2.3 Fill factor (FF)

FF is another significant parameter deficit in kesterite solar cell, which is mainly limited by the series resistance (R_S) in the device. Careful study has been conducted to identify the origin of the R_S, which implies that nonohmic back contact contributes greatly [137]. The nonohmic back contact could be resulted from Schottky barrier formation and/or secondary phase formation at the CZTS/Mo back interface. It is especially severe in high band gap CZTS because of the concentration decrease. Furthermore, the presence of low band gap secondary phases in CZTS could act as shunting pathway, which limits the FF and is therefore undesirable in the device as well. Deliberate investigation and control of the chemical reaction at the back interface have been proposed to modify the back interface microstructure and improve the device performance including FF [138, 139]. Introducing barrier layer at the CZTS/Mo interface has also been attempted in improving the back interface quality [140, 141].

5. Conclusion

Kesterite CZTS material is an emerging and promising PV technology, which could have the opportunity to realize low-cost and high-volume thin film solar cell production. After undergoing a rapid development in last decades, the technology has attracted increasing attention and demonstrated high potential in high efficiency. Experimental and theoretical studies reveal that the quaternary material exits in kesterite structure, similar to the chalcogenide $CuInS_2$. It has a narrow stable region regarding the chemical potential, which makes the formation of secondary phases very easy. The intrinsic defects characteristic in the CZTS is complex where the deep level antisites defects and defect cluster could be prevalent. These unique properties make it struggling in achieving high-efficiency kesterite solar cells. The sensitive defects environment causes undesirable band tailing, electrostatic potential fluctuations, etc., which become recombination path and limit the open-circuit voltage, as well as other PV performance. The formation of secondary phases can lead to high serious resistance, shunting path depends on the band gap of the phase, which will also limit the device performance. In addition, the heterojunction interface and the Mo/CZTS back interface could also contribute to the performance loss because of the unfavorable band alignment, unpassivated defects, or detrimental reaction during high-temperature annealing. The solutions and approaches for tackling these loss mechanisms have been reviewed at the end of the chapter. Finally, in order to build a successful kesterite CZTS technology, combining the advanced approaches summarized in this chapter with further exploring the material synthesis and device physics would pave a path for higher efficiency.

Acknowledgements

This work was supported by the National Key R&D Program of China (No. 2018YFE0203400), the Science and Technology Innovation Program of Hunan Province (No. 2020RC2005), Australian Renewable Energy Agency (ARENA, 2017/RND006), and International Postdoctoral Exchange Fellowship Program (YJ20200207). X.H. acknowledges Australian Research Council (ARC) Future Fellowship (FT190100756). K.S. acknowledges the support from Australian Government through the Australian Renewable Energy Agency (ARENA) and the Australian Centre of Advanced Photovoltaics (ACAP, Grant No.1-SRI001).

Conflict of interest

The authors declare no conflict of interest.

Author details

Kaiwen Sun[1,2*], Fangyang Liu[1*] and Xiaojing Hao[2*]

1 School of Metallurgy and Environment, Central South University, Changsha, China

2 Australian Centre for Advanced Photovoltaics, School of Photovoltaic and Renewable Energy Engineering, University of New South Wales, Sydney, Australia

*Address all correspondence to: kaiwen.sun@unsw.edu.au, liufangyang@csu.edu.cn and xj.hao@unsw.edu.au

References

[1] Nakamura M, Yamaguchi K, Kimoto Y, Yasaki Y, Kato T, Sugimoto H. Cd-Free Cu(In,Ga)$(Se,S)_2$ thin-film solar cell with record efficiency of 23.35%. IEEE Journal of Photovoltaics. 2019;**9**(6):1863-1867. DOI: 10.1109/JPHOTOV.2019.2937218

[2] Green MA, Dunlop ED, Hohl-Ebinger J, Yoshita M, Kopidakis N, Hao X. Solar cell efficiency tables (Version 58). Progress in Photovoltaics: Research and Applications. 2021;**29**(7):657-667. DOI: 10.1002/pip.3444

[3] Commission E. Restriction of Hazardous Substances in Electrical and Electronic Equipment (RoHS). Strasbourg: European Commission; 2011 Available from: https://ec.europa.eu/environment/topics/waste-and-recycling/rohs-directive_en#ecl-inpage-622

[4] Commission E. Critical Raw Materials. Brussels: European Commission; 2011 Available from: https://ec.europa.eu/growth/sectors/raw-materials/areas-specific-interest/critical-raw-materials_en

[5] Walsh A, Chen S, Wei S-H, Gong X-G. Kesterite thin-film solar cells: Advances in materials modelling of Cu2ZnSnS4. Advanced Energy Materials. 2012;**2**(4):400-409. DOI: 10.1002/aenm.201100630

[6] Zhou H, Hsu W-C, Duan H-S, Bob B, Yang W, Song T-B, et al. CZTS nanocrystals: A promising approach for next generation thin film photovoltaics. Energy & Environmental Science. 2013;**6**(10):2822-2838. DOI: 10.1039/C3EE41627E

[7] Ito K. An overview of CZTS-based thin-film solar cells. In: Copper Zinc Tin Sulfide-Based Thin-Film Solar Cells. West Sussex, United Kingdom: John Wiley & Sons, Ltd.; 2014. pp. 1-41

[8] Katagiri H, Jimbo K, Maw WS, Oishi K, Yamazaki M, Araki H, et al. Development of CZTS-based thin film solar cells. Thin Solid Films. 2009;**517**(7):2455-2460. DOI: 10.1016/j.tsf.2008.11.002

[9] Mitzi DB, Gunawan O, Todorov TK, Wang K, Guha S. The path towards a high-performance solution-processed kesterite solar cell. Solar Energy Materials and Solar Cells. 2011;**95**(6):1421-1436. DOI: 10.1016/j.solmat.2010.11.028

[10] Yan C, Huang J, Sun K, Johnston S, Zhang Y, Sun H, et al. Cu2ZnSnS4 solar cells with over 10% power conversion efficiency enabled by heterojunction heat treatment. Nature Energy. 2018;**3**(9):764-772. DOI: 10.1038/s41560-018-0206-0

[11] Li J, Huang Y, Huang J, Liang G, Zhang Y, Rey G, et al. Defect control for 12.5% efficiency Cu2ZnSnSe4 Kesterite thin-film solar cells by Engineering of local chemical environment. Advanced Materials. 2020;**32**(52):2005268. DOI: 10.1002/adma.202005268

[12] NREL. Best Research-Cell Efficiency Chart. Golden, United States: The National Renewable Energy Laboratory; 2021 Available from: https://www.nrel.gov/pv/cell-efficiency.html.

[13] Nitsche R, Sargent DF, Wild P. Crystal growth of quaternary 122464 chalcogenides by iodine vapor transport. Journal of Crystal Growth. 1967;**1**(1):52-53. DOI: 10.1016/0022-0248(67)90009-7

[14] Ito K, Nakazawa T. Electrical and optical properties of Stannite-type quaternary semiconductor thin films. Japanese Journal of Applied Physics. 1988;**27**(Part 1, No 11):2094-2097. DOI: 10.1143/jjap.27.2094

[15] Ito K, Nakazawa T. Stannite-type photovoltaic thin films. In: Proceedings

of the 4th International Conference of Photovoltaic Science and Engineering. Sydney, NSW Australia: Institution of Radio and Electronics Engineers Australia; 1989

[16] Katagiri H, Sasaguchi N, Hando S, Hoshino S, Ohashi J, Yokota T. Preparation and evaluation of Cu_2ZnSnS_4 thin films by sulfurization of E-B evaporated precursor. In: Technical Digest of the 9th International Conference of Photovoltaic Science and Engineering. Miyazaki, Japan: PVSEC-9; 1996

[17] Katagiri H, Sasaguchi N, Hando S, Hoshino S, Ohashi J, Yokota T. Preparation and evaluation of Cu2ZnSnS4 thin films by sulfurization of E·B evaporated precursors. Solar Energy Materials and Solar Cells. 1997;**49**(1):407-414. DOI: 10.1016/S0927-0248(97)00119-0

[18] Friedlmeier TM, Wieser N, Walter T, Dittrich H, Schock H. Heterojunctions based on Cu2ZnSnS4 and Cu2ZnSnSe4 thin films. In: Proceedings of the 14th European Conference of Photovoltaic Science and Engineering and Exhibition. Barcelona, Spain: European Photovoltaic Solar Energy Conference; 1997

[19] Katagiri H. Cu2ZnSnS4 thin film solar cells. Thin Solid Films. 2005; **480-481**:426-432. DOI: 10.1016/j.tsf.2004.11.024

[20] Katagiri H, Ishigaki N, Ishida T, Saito K. Characterization of Cu2ZnSnS4 thin films prepared by vapor phase sulfurization. Japanese Journal of Applied Physics. 2001;**40**(Part 1, No. 2A):500-504.DOI: 10.1143/jjap.40.500

[21] Katagiri H, Saitoh K, Washio T, Shinohara H, Kurumadani T, Miyajima S. Development of thin film solar cell based on Cu2ZnSnS4 thin films. Solar Energy Materials and Solar Cells. 2001;**65**(1):141-148. DOI: 10.1016/S0927-0248(00)00088-X

[22] Katagiri H, Jimbo K, Moriya K, Tsuchida K, editors. Solar cell without environmental pollution by using CZTS thin film. 3rd World Conference on Photovoltaic Energy Conversion. Osaka, Japan: IEEE; 2003

[23] Kobayashi T, Jimbo K, Tsuchida K, Shinoda S, Oyanagi T, Katagiri H. Investigation of Cu2ZnSnS4-based thin film solar cells using abundant materials. Japanese Journal of Applied Physics. 2005;**44**(1B):783-787. DOI: 10.1143/jjap.44.783

[24] Jimbo K, Kimura R, Kamimura T, Yamada S, Maw WS, Araki H, et al. Cu2ZnSnS4-type thin film solar cells using abundant materials. Thin Solid Films. 2007;**515**(15):5997-5999. DOI: 10.1016/j.tsf.2006.12.103

[25] Katagiri H, Jimbo K, Yamada S, Kamimura T, Maw WS, Fukano T, et al. Enhanced conversion efficiencies of Cu2ZnSnS4-based thin film solar cells by using preferential etching technique. Applied Physics Express. 2008;**1**:041201. DOI: 10.1143/apex.1.041201

[26] Tajima S, Itoh T, Hazama H, Ohishi K, Asahi R. Improvement of the open-circuit voltage of Cu_2ZnSnS_4 cells using a two-layered process. In: IEEE 40th Photovoltaic Specialist Conference (PVSC). Denver, CO, USA: IEEE; 2014. DOI: 10.1109/PVSC.2014.6924951

[27] Todorov TK, Reuter KB, Mitzi DB. High-efficiency solar cell with Earth-abundant liquid-processed absorber. Advanced Materials. 2010;**22**(20):E156-E1E9. DOI: 10.1002/adma.200904155

[28] Repins I, Beall C, Vora N, DeHart C, Kuciauskas D, Dippo P, et al. Co-evaporated Cu2ZnSnSe4 films and devices. Solar Energy Materials and Solar Cells. 2012;**101**:154-159. DOI: 10.1016/j.solmat.2012.01.008

[29] Sugimoto H, Liao C, Hiroi H, Sakai N, Kato T. Lifetime improvement

for high efficiency Cu_2ZnSnS_4 submodules. In: IEEE 39th Photovoltaic Specialists Conference (PVSC). Tampa, FL, USA: IEEE; 2013. DOI: 10.1109/PVSC.2013.6745135

[30] Haass SG, Diethelm M, Werner M, Bissig B, Romanyuk YE, Tiwari AN. 11.2% Efficient solution processed Kesterite solar cell with a low voltage deficit. Advanced Energy Materials. 2015;**5**(18):1500712(1-7). DOI: 10.1002/aenm.201500712

[31] Lin X, Kavalakkatt J, Lux-Steiner MC, Ennaoui A. Inkjet-printed Cu2ZnSn(S, Se)4 solar cells. Advanced Science. 2015;**2**:1500028(1-6). DOI: 10.1002/advs.201500028

[32] Schnabel T, Abzieher T, Friedlmeier TM, Ahlswede E. Solution-based preparation of Cu2ZnSn(S,Se)4 for solar cells—Comparison of SnSe2 and elemental Se as Chalcogen source. IEEE Journal of Photovoltaics. 2015;**5**(2):670-675. DOI: 10.1109/JPHOTOV.2015.2392935

[33] Sun K, Yan C, Liu F, Huang J, Zhou F, Stride JA, et al. Over 9% efficient Kesterite Cu2ZnSnS4 solar cell fabricated by using Zn1–xCdxS buffer layer. Advanced Energy Materials. 2016;**6**(12):1600046. DOI: 10.1002/aenm.201600046

[34] Barkhouse DAR, Gunawan O, Gokmen T, Todorov TK, Mitzi DB. Device characteristics of a 10.1% hydrazine-processed Cu2ZnSn(Se,S)4 solar cell. Progress in Photovoltaics: Research and Applications. 2012;**20**(1):6-11. DOI: 10.1002/pip.1160

[35] Qijie G, Yanyan C, Caspar JV, Farneth WE, Ionkin AS, Johnson LK, et al. A simple solution-based route to high-efficiency CZTSSe thin-film solar cells. In: Photovoltaic Specialists Conference (PVSC), 2012 38th IEEE. Austin, TX, USA: IEEE; 2012. DOI: 10.1109/PVSC.2012.6318213

[36] Yang W, Duan H-S, Bob B, Zhou H, Lei B, Chung C-H, et al. Novel solution processing of high-efficiency Earth-abundant Cu2ZnSn(S,Se)4 solar cells. Advanced Materials. 2012;**24**(47):6323-6329. DOI: 10.1002/adma.201201785

[37] Xin H, Katahara JK, Braly IL, Hillhouse HW. 8% Efficient Cu2ZnSn(S,Se)4 solar cells from redox equilibrated simple precursors in DMSO. Advanced Energy Materials. 2014;**4**(11):1301823 (1-5). DOI: 10.1002/aenm.201301823

[38] Shin B, Gunawan O, Zhu Y, Bojarczuk NA, Chey SJ, Guha S. Thin film solar cell with 8.4% power conversion efficiency using an earth-abundant Cu2ZnSnS4 absorber. Progress in Photovoltaics: Research and Applications. 2013;**21**(1):72-76. DOI: 10.1002/pip.1174

[39] Wang W, Winkler MT, Gunawan O, Gokmen T, Todorov TK, Zhu Y, et al. Device characteristics of CZTSSe thin-film solar cells with 12.6% efficiency. Advanced Energy Materials. 2014;**4**(7): 1301465 (1-5). DOI: 10.1002/aenm.201301465

[40] Ahmed S, Reuter KB, Gunawan O, Guo L, Romankiw LT, Deligianni H. A high efficiency electrodeposited Cu2ZnSnS4 solar cell. Advanced Energy Materials. 2012;**2**(2):253-259. DOI: 10.1002/aenm.201100526

[41] Todorov TK, Tang J, Bag S, Gunawan O, Gokmen T, Zhu Y, et al. Beyond 11% efficiency: Characteristics of state-of-the-art Cu2ZnSn(S,Se)4 solar cells. Advanced Energy Materials. 2012;**2013**:34-38. DOI: 10.1002/aenm.201200348

[42] Yang K-J, Son D-H, Sung S-J, Sim J-H, Kim Y-I, Park S-N, et al. A band-gap-graded CZTSSe solar cell with 12.3% efficiency. Journal of Materials Chemistry A. 2016;**4**(26):10151-10158. DOI: 10.1039/C6TA01558A

[43] Son D-H, Kim S-H, Kim S-Y, Kim Y-I, Sim J-H, Park S-N, et al. Effect of solid-H2S gas reactions on CZTSSe thin film growth and photovoltaic properties of a 12.62% efficiency device. Journal of Materials Chemistry A. 2019;7(44): 25279-25289. DOI: 10.1039/C9TA08310C

[44] Su Z, Liang G, Fan P, Luo J, Zheng Z, Xie Z, et al. Device postannealing enabling over 12% efficient solution-processed Cu2ZnSnS4 solar cells with Cd2+ substitution. Advanced Materials. 2020;**32**(32): 2000121. DOI: 10.1002/adma.202000121

[45] Du Y, Wang S, Tian Q, Zhao Y, Chang X, Xiao H, et al. Defect Engineering in Earth-abundant Cu2ZnSn(S,Se)4 photovoltaic materials via Ga3+-doping for over 12% efficient solar cells. Advanced Functional Materials. 2021;**31**(16):2010325. DOI: 10.1002/adfm.202010325

[46] Gong Y, Qiu R, Niu C, Fu J, Jedlicka E, Giridharagopal R, et al. Ag Incorporation with controlled grain growth enables 12.5% efficient Kesterite solar cell with open circuit voltage reached 64.2% Shockley–Queisser limit. Advanced Functional Materials. 2021;**31**(24):2101927. DOI: 10.1002/ adfm.202101927

[47] Chen S, Gong XG, Walsh A, Wei S-H. Electronic structure and stability of quaternary chalcogenide semiconductors derived from cation cross-substitution of II-VI and I-III-VI compounds. Physical Review B. 2009;**79**(16):165211. DOI: 10.1103/ PhysRevB.79.165211

[48] Chen S, Gong XG, Walsh A, Wei S-H. Crystal and electronic band structure of Cu2ZnSnX4 (X=S and Se) photovoltaic absorbers: First-principles insights. Applied Physics Letters. 2009;**94**(4):041903. DOI: 10.1063/1. 3074499

[49] Hall S, Szymanski J, Stewart J. Kesterite, (Cu< 2)(Zn, Fe) SnS< 4), and stannite, Cu< 2)(Fe, Zn) SnS< 4), structurally similar but distinct minerals. The Canadian Mineralogist. 1978;**16**(2):131-137

[50] Schorr S, Hoebler H-J, Tovar M. A neutron diffraction study of the stannite-kesterite solid solution series. European Journal of Mineralogy. 2007;**19**(1):65-73. DOI: 10.1127/0935-1221/2007/0019-0065

[51] Friedelmeier T, Dittrich H, Schock H-W. Growth and characterization of Cu2ZnSnS4 and Cu2ZnSnSe4 thin films for photovoltaic applications. In: Ternary and Multinary Compounds. University of Salford, UK: CRC Press; 1998. p. 345

[52] Bernardini GP, Borrini D, Caneschi A, Di Benedetto F, Gatteschi D, Ristori S, et al. EPR and SQUID magnetometry study of Cu2FeSnS4 (stannite) and Cu2ZnSnS4 (kesterite). Physics and Chemistry of Minerals. 2000;**27**(7):453-461. DOI: 10.1007/s002690000086

[53] Schorr S. The crystal structure of kesterite type compounds: A neutron and X-ray diffraction study. Solar Energy Materials and Solar Cells. 2011;**95**(6):1482-1488. DOI: 10.1016/j. solmat.2011.01.002

[54] Hall S, Kissin S, Stewart J. Stannite and kesterite-distinct minerals or components of a solid-solution. In: Acta crystallographica section A. Copenhagen, Denmark: Munksgaard Int Publ Ltd.; Vol. 31. 1975 p. s67

[55] Seol J-S, Lee S-Y, Lee J-C, Nam H-D, Kim K-H. Electrical and optical properties of Cu2ZnSnS4 thin films prepared by rf magnetron sputtering process. Solar Energy Materials and Solar Cells. 2003;**75**(1):155-162. DOI: 10.1016/S0927-0248(02)00127-7

[56] Babu GS, Kumar YBK, Bhaskar PU, Raja VS. Effect of post-deposition annealing on the growth of

Cu2ZnSnSe4 thin films for a solar cell absorber layer. Semiconductor Science and Technology. 2008;**23**(8):085023. DOI: 10.1088/0268-1242/23/8/085023

[57] Chen S, Walsh A, Luo Y, Yang J-H, Gong XG, Wei S-H. Wurtzite-derived polytypes of kesterite and stannite quaternary chalcogenide semiconductors. Physical Review B. 2010;**82**(19):195203. DOI: 10.1103/PhysRevB.82.195203

[58] Paier J, Asahi R, Nagoya A, Kresse G. Cu_2ZnSnS_4 as a potential photovoltaic material: A hybrid Hartree-Fock density functional theory study. Physical Review B. 2009;**79**(11):115126. DOI: 10.1103/PhysRevB.79.115126

[59] Persson C. Electronic and optical properties of Cu2ZnSnS4 and Cu2ZnSnSe4. Journal of Applied Physics. 2010;**107**(5):053710. DOI: 10.1063/1.3318468

[60] Schorr S. Structural aspects of adamantine like multinary chalcogenides. Thin Solid Films. 2007;**515**(15):5985-5991. DOI: 10.1016/j.tsf.2006.12.100

[61] Botti S, Kammerlander D, Marques MAL. Band structures of Cu2ZnSnS4 and Cu2ZnSnSe4 from many-body methods. Applied Physics Letters. 2011;**98**(24):241915. DOI: 10.1063/1.3600060

[62] Ichimura M, Nakashima Y. Analysis of atomic and electronic structures of Cu2ZnSnS4 based on first-principle calculation. Japanese Journal of Applied Physics. 2009;**48**(9):090202. DOI: 10.1143/jjap.48.090202

[63] Persson C, Chen R, Zhao H, Kumar M, Huang D. Electronic Structure and Optical Properties from First-Principles Modeling. In: Copper Zinc Tin Sulfide-Based Thin-Film Solar Cells. West Sussex, United Kingdom: John Wiley & Sons, Ltd.; 2014. pp. 75-105

[64] Nakayama N, Ito K. Sprayed films of stannite Cu2ZnSnS4. Applied Surface Science. 1996;**92**:171-175. DOI: 10.1016/0169-4332(95)00225-1

[65] Kamoun N, Bouzouita H, Rezig B. Fabrication and characterization of Cu2ZnSnS4 thin films deposited by spray pyrolysis technique. Thin Solid Films. 2007;**515**(15):5949-5952. DOI: 10.1016/j.tsf.2006.12.144

[66] Gao F, Yamazoe S, Maeda T, Nakanishi K, Wada T. Structural and optical properties of In-free Cu2ZnSn(S,Se)4 solar cell materials. Japanese Journal of Applied Physics. 2012;**51**:10NC29. DOI: 10.1143/jjap.51.10nc29

[67] Tanaka K, Fukui Y, Moritake N, Uchiki H. Chemical composition dependence of morphological and optical properties of Cu2ZnSnS4 thin films deposited by sol–gel sulfurization and Cu2ZnSnS4 thin film solar cell efficiency. Solar Energy Materials and Solar Cells. 2011;**95**(3):838-842. DOI: 10.1016/j.solmat.2010.10.031

[68] Patel M, Mukhopadhyay I, Ray A. Structural, optical and electrical properties of spray-deposited CZTS thin films under a non-equilibrium growth condition. Journal of Physics D: Applied Physics. 2012;**45**(44):445103. DOI: 10.1088/0022-3727/45/44/445103

[69] Ahn S, Jung S, Gwak J, Cho A, Shin K, Yoon K, et al. Determination of band gap energy (Eg) of Cu2ZnSnSe4 thin films: On the discrepancies of reported band gap values. Applied Physics Letters. 2010;**97**(2):021905. DOI: 10.1063/1.3457172

[70] Haight R, Barkhouse A, Gunawan O, Shin B, Copel M, Hopstaken M, et al. Band alignment at the Cu2ZnSn(SxSe1−x)4/CdS interface. Applied Physics Letters. 2011;**98**(25):253502. DOI: 10.1063/1.3600776

[71] He J, Sun L, Chen S, Chen Y, Yang P, Chu J. Composition dependence of

structure and optical properties of Cu2ZnSn(S,Se)4 solid solutions: An experimental study. Journal of Alloys and Compounds. 2012;**511**(1):129-132. DOI: 10.1016/j.jallcom.2011.08.099

[72] Ennaoui A, Lux-Steiner M, Weber A, Abou-Ras D, Kötschau I, Schock HW, et al. Cu2ZnSnS4 thin film solar cells from electroplated precursors: Novel low-cost perspective. Thin Solid Films. 2009;**517**(7):2511-2514. DOI: 10.1016/j.tsf.2008.11.061

[73] Kumar M, Zhao H, Persson C. Cation vacancies in the alloy compounds of Cu2ZnSn(S1−xSex)4 and CuIn(S1−xSex)2. Thin Solid Films. 2013;**535**:318-321. DOI: 10.1016/j.tsf.2012.11.063

[74] Chen S, Walsh A, Gong X-G, Wei S-H. Classification of lattice defects in the Kesterite Cu2ZnSnS4 and Cu2ZnSnSe4 Earth-abundant solar cell absorbers. Advanced Materials. 2013;**25**(11):1522-1539. DOI: 10.1002/adma.201203146

[75] Lund EA, Du H, Hlaing Oo WM, Teeter G, Scarpulla MA. Investigation of combinatorial coevaporated thin film Cu2ZnSnS4 (II): Beneficial cation arrangement in Cu-rich growth. Journal of Applied Physics. 2014;**115**(17):173503. DOI: 10.1063/1.4871665

[76] Chen S, Gong XG, Walsh A, Wei S-H. Defect physics of the kesterite thin-film solar cell absorber Cu2ZnSnS4. Applied Physics Letters. 2010;**96**(2):021902. DOI: 10.1063/1.3275796

[77] Olekseyuk ID, Dudchak IV, Piskach LV. Phase equilibria in the Cu2S–ZnS–SnS2 system. Journal of Alloys and Compounds. 2004;**368**(1):135-143. DOI: 10.1016/j.jallcom.2003.08.084

[78] Dudchak IV, Piskach LV. Phase equilibria in the Cu2SnSe3–SnSe2–ZnSe system. Journal of Alloys and Compounds. 2003;**351**(1):145-150. DOI: 10.1016/S0925-8388(02)01024-1

[79] Chen S, Yang J-H, Gong XG, Walsh A, Wei S-H. Intrinsic point defects and complexes in the quaternary kesterite semiconductor Cu_2ZnSnS_4. Physical Review B. 2010;**81**(24):245204. DOI: 10.1103/PhysRevB.81.245204

[80] Siebentritt S, Schorr S. Kesterites—a challenging material for solar cells. Progress in Photovoltaics: Research and Applications. 2012;**20**(5): 512-519. DOI: 10.1002/pip.2156

[81] Sun K, Huang J, Yan C, Pu A, Liu F, Sun H, et al. Self-assembled nanometer-scale ZnS structure at the CZTS/ZnCdS heterointerface for high-efficiency wide band gap Cu2ZnSnS4 solar cells. Chemistry of Materials. 2018;**30**(12): 4008-4016. DOI: 10.1021/acs.chemmater.8b00009

[82] Berg DM, Djemour R, Gütay L, Zoppi G, Siebentritt S, Dale PJ. Thin film solar cells based on the ternary compound Cu2SnS3. Thin Solid Films. 2012;**520**(19):6291-6294. DOI: 10.1016/j.tsf.2012.05.085

[83] Marcano G, Rincón C, de Chalbaud LM, Bracho DB, Pérez GS. Crystal growth and structure, electrical, and optical characterization of the semiconductor Cu2SnSe3. Journal of Applied Physics. 2001;**90**(4):1847-1853. DOI: 10.1063/1.1383984

[84] Yu P, Cardona M. Fundamentals of Semiconductors. 4th ed. Berlin, Heidelberg, Germany: Springer; 2010

[85] Lin Y-T, Shi J-B, Chen Y-C, Chen C-J, Wu P-F. Synthesis and characterization of Tin Disulfide (SnS2) nanowires. Nanoscale Research Letters. 2009;**4**(7):694. DOI: 10.1007/s11671-009-9299-5

[86] Sava F, Lörinczi A, Popescu M, Socol G, Axente E, Mihailescu IN, et al. Amorphous SnSe2 films. Journal of Optoelectronics and Advanced Materials. 2006;**8**(4):1367-1371

[87] Vidal J, Lany S, d'Avezac M, Zunger A, Zakutayev A, Francis J, et al. Band-structure, optical properties, and defect physics of the photovoltaic semiconductor SnS. Applied Physics Letters. 2012;**100**(3):032104. DOI: 10.1063/1.3675880

[88] Sinsermsuksakul P, Heo J, Noh W, Hock AS, Gordon RG. Atomic layer deposition of Tin monosulfide thin films. Advanced Energy Materials. 2011;**1**(6):1116-1125. DOI: 10.1002/aenm.201100330

[89] Franzman MA, Schlenker CW, Thompson ME, Brutchey RL. Solution-phase synthesis of SnSe nanocrystals for use in solar cells. Journal of the American Chemical Society. 2010;**132**(12): 4060-4061. DOI: 10.1021/ja100249m

[90] Liu G, Schulmeyer T, Brötz J, Klein A, Jaegermann W. Interface properties and band alignment of Cu2S/CdS thin film solar cells. Thin Solid Films. 2003;**431-432**:477-482. DOI: 10.1016/S0040-6090(03)00190-1

[91] Kashida S, Shimosaka W, Mori M, Yoshimura D. Valence band photoemission study of the copper chalcogenide compounds, Cu2S, Cu2Se and Cu2Te. Journal of Physics and Chemistry of Solids. 2003;**12**:2357-2363. DOI: 10.1016/S0022-3697(03)00272-5

[92] Siebentritt S. Why are kesterite solar cells not 20% efficient? Thin Solid Films. 2013;**535**:1-4. DOI: 10.1016/j.tsf.2012.12.089

[93] Timo Wätjen J, Engman J, Edoff M, Platzer-Björkman C. Direct evidence of current blocking by ZnSe in Cu2ZnSnSe4 solar cells. Applied Physics Letters. 2012;**100**(17):173510. DOI: 10.1063/1.4706256

[94] Redinger A, Mousel M, Wolter MH, Valle N, Siebentritt S. Influence of S/Se ratio on series resistance and on dominant recombination pathway in Cu2ZnSn(SSe)4 thin film solar cells. Thin Solid Films. 2013;**535**:291-295. DOI: 10.1016/j.tsf.2012.11.111

[95] Mendis BG, Goodman MCJ, Major JD, Taylor AA, Durose K, Halliday DP. The role of secondary phase precipitation on grain boundary electrical activity in Cu2ZnSnS4 (CZTS) photovoltaic absorber layer material. Journal of Applied Physics. 2012;**112** (12):124508. DOI: 10.1063/1.4769738

[96] Sun K, Yan C, Huang J, Liu F, Li J, Sun H, et al. Beyond 10% efficiency Cu2ZnSnS4 solar cells enabled by modifying the heterojunction interface chemistry. Journal of Materials Chemistry A. 2019;**7**(48):27289-27296. DOI: 10.1039/C9TA09576D

[97] Redinger A, Siebentritt S. Coevaporation of Cu2ZnSnSe4 thin films. Applied Physics Letters. 2010;**97**(9):092111. DOI: 10.1063/1.3483760

[98] Mousel M, Redinger A, Djemour R, Arasimowicz M, Valle N, Dale P, et al. HCl and Br2-MeOH etching of Cu2ZnSnSe4 polycrystalline absorbers. Thin Solid Films. 2013;**535**:83-87. DOI: 10.1016/j.tsf.2012.12.095

[99] Nagoya A, Asahi R, Wahl R, Kresse G. Defect formation and phase stability of Cu2ZnSnS4 photovoltaic material. Physical Review B. 2010;**81**(11):113202. DOI: 10.1103/PhysRevB.81.113202

[100] Wibowo RA, Kim WS, Lee ES, Munir B, Kim KH. Single step preparation of quaternary Cu2ZnSnSe4 thin films by RF magnetron sputtering from binary chalcogenide targets. Journal of Physics and Chemistry of Solids. 2007;**68**(10):1908-1913. DOI: 10.1016/j.jpcs.2007.05.022

[101] Zhang X, Shi X, Ye W, Ma C, Wang C. Electrochemical deposition of quaternary Cu2ZnSnS4 thin films as potential solar cell material. Applied Physics A. 2009;**94**(2):381-386. DOI: 10.1007/s00339-008-4815-5

[102] Tanaka T, Nagatomo T, Kawasaki D, Nishio M, Guo Q, Wakahara A, et al. Preparation of Cu2ZnSnS4 thin films by hybrid sputtering. Journal of Physics and Chemistry of Solids. 2005;**66**(11):1978-1981. DOI: 10.1016/j.jpcs.2005.09.037

[103] Scragg JJ, Dale PJ, Peter LM, Zoppi G, Forbes I. New routes to sustainable photovoltaics: evaluation of Cu2ZnSnS4 as an alternative absorber material. Physica Status Solidi (b). 2008;**245**(9):1772-1778. DOI: 10.1002/pssb.200879539

[104] Altosaar M, Raudoja J, Timmo K, Danilson M, Grossberg M, Krustok J, et al. Cu2Zn1–x Cdx Sn(Se1–y Sy)4 solid solutions as absorber materials for solar cells. Physica Status Solidi (a). 2008;**205**(1):167-170. DOI: 10.1002/pssa.200776839

[105] Oishi K, Saito G, Ebina K, Nagahashi M, Jimbo K, Maw WS, et al. Growth of Cu2ZnSnS4 thin films on Si (100) substrates by multisource evaporation. Thin Solid Films. 2008;**517**(4):1449-1452. DOI: 10.1016/j.tsf.2008.09.056

[106] Kishore Kumar YB, Suresh Babu G, Uday Bhaskar P, Sundara RV. Preparation and characterization of spray-deposited Cu2ZnSnS4 thin films. Solar Energy Materials and Solar Cells. 2009;**93**(8):1230-1237. DOI: 10.1016/j.solmat.2009.01.011

[107] Hönes K, Zscherpel E, Scragg J, Siebentritt S. Shallow defects in Cu2ZnSnS4. Physica B: Condensed Matter. 2009;**404**(23):4949-4952. DOI: 10.1016/j.physb.2009.08.206

[108] Shinde NM, Dubal DP, Dhawale DS, Lokhande CD, Kim JH, Moon JH. Room temperature novel chemical synthesis of Cu2ZnSnS4 (CZTS) absorbing layer for photovoltaic application. Materials Research Bulletin. 2012;**47**(2):302-307. DOI: 10.1016/j.materresbull.2011.11.020

[109] Miyamoto Y, Tanaka K, Oonuki M, Moritake N, Uchiki H. Optical properties of Cu2ZnSnS4 thin films prepared by Sol–Gel and sulfurization method. Japanese Journal of Applied Physics. 2008;**47**(1):596-597. DOI: 10.1143/jjap.47.596

[110] Prabhakar T, Jampana N. Effect of sodium diffusion on the structural and electrical properties of Cu2ZnSnS4 thin films. Solar Energy Materials and Solar Cells. 2011;**95**(3):1001-1004. DOI: 10.1016/j.solmat.2010.12.012

[111] Cui X, Sun K, Huang J, Yun JS, Lee C-Y, Yan C, et al. Cd-Free Cu2ZnSnS4 solar cell with an efficiency greater than 10% enabled by Al2O3 passivation layers. Energy & Environmental Science. 2019;**12**(9):2751-2764. DOI: 10.1039/C9EE01726G

[112] Cazzaniga A, Crovetto A, Yan C, Sun K, Hao X, Ramis Estelrich J, et al. Ultra-thin Cu2ZnSnS4 solar cell by pulsed laser deposition. Solar Energy Materials and Solar Cells. 2017;**166**:91-99. DOI: 10.1016/j.solmat.2017.03.002

[113] Guo Q, Ford GM, Yang W-C, Walker BC, Stach EA, Hillhouse HW, et al. Fabrication of 7.2% Efficient CZTSSe Solar Cells Using CZTS Nanocrystals. Journal of the American Chemical Society. 2010;**132**(49):17384-17386. DOI: 10.1021/ja108427b

[114] Cao Y, Denny MS, Caspar JV, Farneth WE, Guo Q, Ionkin AS, et al. High-efficiency solution-processed Cu2ZnSn(S,Se)4 thin-film solar cells prepared from binary and ternary nanoparticles. Journal of the American Chemical Society. 2012;**134**(38):15644-15647. DOI: 10.1021/ja3057985

[115] Qi Y-F, Kou D-X, Zhou W-H, Zhou Z-J, Tian Q-W, Meng Y-N, et al. Engineering of interface band bending and defects elimination via a Ag-graded active layer for efficient

(Cu,Ag)2ZnSn(S,Se)4 solar cells. Energy & Environmental Science. 2017;**10**(11):2401-2410. DOI: 10.1039/C7EE01405H

[116] Wangperawong A, King JS, Herron SM, Tran BP, Pangan-Okimoto K, Bent SF. Aqueous bath process for deposition of Cu2ZnSnS4 photovoltaic absorbers. Thin Solid Films. 2011;**519**(8):2488-2492. DOI: 10.1016/j.tsf.2010.11.040

[117] Sun K, Wang A, Su Z, Liu F, Hao X. Enhancing the performance of Cu2ZnSnS4 solar cell fabricated via successive ionic layer adsorption and reaction method by optimizing the annealing process. Solar Energy. 2021;**220**:204-210. DOI: 10.1016/j.solener.2021.03.033

[118] Zhou Z, Wang Y, Xu D, Zhang Y. Fabrication of Cu2ZnSnS4 screen printed layers for solar cells. Solar Energy Materials and Solar Cells. 2010;**94**(12):2042-2045. DOI: 10.1016/j.solmat.2010.06.010

[119] Winkler MT, Wang W, Gunawan O, Hovel HJ, Todorov TK, Mitzi DB. Optical designs that improve the efficiency of Cu2ZnSn(S,Se)4 solar cells. Energy & Environmental Science. 2014;7(3):1029-1036. DOI: 10.1039/C3EE42541J

[120] Shockley W, Queisser HJ. Detailed balance limit of efficiency of p-n junction solar cells. Journal of Applied Physics. 1961;**32**(3):510-519. DOI: 10.1063/1.1736034

[121] Gokmen T, Gunawan O, Todorov TK, Mitzi DB. Band tailing and efficiency limitation in kesterite solar cells. Applied Physics Letters. 2013;**103**(10):103506. DOI: 10.1063/1.4820250

[122] Chirilă A, Buecheler S, Pianezzi F, Bloesch P, Gretener C, Uhl AR, et al. Highly efficient Cu(In,Ga)Se2 solar cells grown on flexible polymer films. Nature Materials. 2011;**10**(11):857-861. DOI: 10.1038/nmat3122

[123] Su Z, Tan JMR, Li X, Zeng X, Batabyal SK, Wong LH. Cation substitution of solution-processed Cu2ZnSnS4 thin film solar cell with over 9% efficiency. Advanced Energy Materials. 2015;**5**(19):1500682. DOI: 10.1002/aenm.201500682

[124] Yan C, Sun K, Huang J, Johnston S, Liu F, Veettil BP, et al. Beyond 11% efficient sulfide Kesterite Cu2ZnxCd1–xSnS4 solar cell: Effects of Cadmium alloying. ACS Energy Letters. 2017;**2**(4):930-936. DOI: 10.1021/acsenergylett.7b00129

[125] Hages CJ, Levcenco S, Miskin CK, Alsmeier JH, Abou-Ras D, Wilks RG, et al. Improved performance of Ge-alloyed CZTGeSSe thin-film solar cells through control of elemental losses. Progress in Photovoltaics: Research and Applications. 2015;**23**(3):376-384. DOI: 10.1002/pip.2442

[126] Rey G, Redinger A, Sendler J, Weiss TP, Thevenin M, Guennou M, et al. The band gap of Cu2ZnSnSe4: Effect of order-disorder. Applied Physics Letters. 2014;**105**(11):112106. DOI: 10.1063/1.4896315

[127] Scragg JJS, Choubrac L, Lafond A, Ericson T, Platzer-Björkman C. A low-temperature order-disorder transition in Cu2ZnSnS4 thin films. Applied Physics Letters. 2014;**104**(4):041911. DOI: 10.1063/1.4863685

[128] Gong Y, Zhang Y, Zhu Q, Zhou Y, Qiu R, Niu C, et al. Identifying the origin of the Voc deficit of kesterite solar cells from the two grain growth mechanisms induced by Sn2+ and Sn4+ precursors in DMSO solution. Energy & Environmental Science. 2021;**14**(4):2369-2380. DOI: 10.1039/D0EE03702H

[129] Yang K-J, Kim S, Kim S-Y, Ahn K, Son D-H, Kim S-H, et al. Flexible Cu2ZnSn(S,Se)4 solar cells with over 10% efficiency and methods of enlarging the cell area. Nature

Communications. 2019;**10**(1):2959. DOI: 10.1038/s41467-019-10890-x

[130] Ericson T, Scragg JJ, Hultqvist A, Wätjen JT, Szaniawski P, Törndahl T, et al. Zn(O, S) buffer layers and thickness variations of CdS buffer for Cu2ZnSnS4 solar cells. IEEE Journal of Photovoltaics. 2014;**4**(1):465-469. DOI: 10.1109/JPHOTOV.2013.2283058

[131] Li J, Huang L, Hou J, Wu X, Niu J, Chen G, et al. Effects of substrate orientation and solution movement in chemical bath deposition on Zn(O,S) buffer layer and Cu(In,Ga)Se2 thin film solar cells. Nano Energy. 2019;**58**:427-436. DOI: 10.1016/j.nanoen.2019.01.054

[132] Cui X, Sun K, Huang J, Lee C-Y, Yan C, Sun H, et al. Enhanced heterojunction interface quality to achieve 9.3% efficient Cd-free Cu2ZnSnS4 solar cells using atomic layer deposition ZnSnO buffer layer. Chemistry of Materials. 2018;**30**(21):7860-7871. DOI: 10.1021/acs.chemmater.8b03398

[133] Larsen JK, Larsson F, Törndahl T, Saini N, Riekehr L, Ren Y, et al. Cadmium free Cu2ZnSnS4 solar cells with 9.7% efficiency. Advanced Energy Materials. 2019;**9**(21):1900439. DOI: 10.1002/aenm.201900439

[134] Yuan Z-K, Chen S, Xiang H, Gong X-G, Walsh A, Park J-S, et al. Engineering solar cell absorbers by exploring the band alignment and defect disparity: The case of Cu- and Ag-based Kesterite compounds. Advanced Functional Materials. 2015;**25**(43):6733-6743. DOI: 10.1002/adfm.201502272

[135] Sun H, Sun K, Huang J, Yan C, Liu F, Park J, et al. Efficiency enhancement of Kesterite Cu2ZnSnS4 solar cells via solution-processed ultrathin Tin oxide intermediate layer at absorber/buffer interface. ACS Applied Energy Materials. 2018;**1**(1):154-160. DOI: 10.1021/acsaem.7b00044

[136] Huang TJ, Yin X, Qi G, Gong H. CZTS-based materials and interfaces and their effects on the performance of thin film solar cells. Physica status solidi (RRL) - Rapid Research Letters. 2014;**08**(09):735-762. DOI: 10.1002/pssr.201409219

[137] Tai KF, Gunawan O, Kuwahara M, Chen S, Mhaisalkar SG, Huan CHA, et al. Fill factor losses in Cu2ZnSn(SxSe1−x)4 solar cells: Insights from physical and electrical characterization of devices and exfoliated films. Advanced Energy Materials. 2016;**6**(3):1501609. DOI: 10.1002/aenm.201501609

[138] Scragg JJ, Wätjen JT, Edoff M, Ericson T, Kubart T, Platzer-Björkman C. A detrimental reaction at the molybdenum back contact in Cu2ZnSn(S,Se)4 thin-film solar cells. Journal of the American Chemical Society. 2012;**134**(47):19330-19333. DOI: 10.1021/ja308862n

[139] Scragg JJ, Kubart T, Wätjen JT, Ericson T, Linnarsson MK, Platzer-Björkman C. Effects of back contact instability on Cu2ZnSnS4 devices and processes. Chemistry of Materials. 2013;**25**(15):3162-3171. DOI: 10.1021/cm4015223

[140] Liu F, Sun K, Li W, Yan C, Cui H, Jiang L, et al. Enhancing the Cu2ZnSnS4 solar cell efficiency by back contact modification: Inserting a thin TiB2 intermediate layer at Cu2ZnSnS4/Mo interface. Applied Physics Letters. 2014;**104**(5):051105. DOI: 10.1063/1.4863736

[141] Liu F, Huang J, Sun K, Yan C, Shen Y, Park J, et al. Beyond 8% ultrathin kesterite Cu2ZnSnS4 solar cells by interface reaction route controlling and self-organized nanopattern at the back contact. NPG Asia Materials. 2017;**9**(7):e401. DOI: 10.1038/am.2017.103

Chapter 5

Outdoor Performance of Perovskite Photovoltaic Technology

Esteban Velilla Hernández, Juan Bernardo Cano Quintero, Juan Felipe Montoya, Iván Mora-Seró and Franklin Jaramillo Isaza

Abstract

In the case of emerging photovoltaic technologies such as perovskite, most published works have focused on laboratory-scale cells, indoor conditions and no international standards have been fully established and adopted. Accordingly, this chapter shows a brief introduction on the standards and evaluation methods for perovskite solar minimodules under natural sunlight conditions. Therefore, we propose evaluating the outdoor performance in terms of power, following the international standard IEC 61853–1 to obtain the performance according to the power rating conditions. After some rigorous experimental evaluations, results shown that the maximum power (P_{max}) evolution for the analyzed minimodules could be correlated with one of the three patterns commonly described for degradation processes in the literature, named convex, linear, and concave. These patterns were used to estimate the degradation rate and lifetime (T_{80}). Moreover, ideality factor (n_{ID}) was estimated from the open-circuit voltage (V_{oc}) dependence on irradiance and ambient temperature (outdoor data) to provide physical insight into the recombination mechanism dominating the performance during the exposure. In this context, it was observed that the three different degradation patterns identified for P_{max} can also be identified by n_{ID}. Finally, based on the linear relationship between T_{80} and the time to first reach n_{ID} = 2 (Tn_{ID2}), is demonstrated that n_{ID} analysis could offer important complementary information with important implications for this technology outdoor development, due that the changes in n_{ID} could be correlated with the recombination mechanisms and degradation processes occurring in the device.

Keywords: perovskite solar cell, outdoor performance, solar minimodule, ideality factor, degradation

1. Introduction

Perovskite solar cells (PSCs) have attracted the attention of the scientific community in the last decade evidenced by the impressive number of scientific publications found under the keyword "perovskite & solar cell" in the database SCOPUS[1].

[1] Database SCOPUS, https://www.scopus.com, accessed on 24th June of 2021.

At the time of writing, 19407 documents were retrieved, corresponding most of these works to small area devices measured in laboratory conditions. However, for the commercialization of this photovoltaic technology (PV), the main challenge is the fabrication of large area devices with high temporal stability and processed by scalable techniques, being a prerequisite for real-world application reliable measurement of perovskite technology under outdoor conditions. In addition, given the characteristics of metal halide perovskites (denoted as "perovskites" in this work), it is to be noting that there is not a standardized protocol for the fabrication and performance measurements.

The relevance of outdoor performance is supported by a deep analysis of the scientific literature which reveals that only a small fraction of research articles deals with perovskite solar cells measured in outdoor conditions. For instance, only 100 documents were found in SCOPUS when "outdoor" was added as keyword in the search. Thus, less than 1% of the documents reported outdoor measurements, illustrating the wide gap between the development of larger area devices and laboratory scale devices (active area, A < = 1 cm^2). Moreover, from the data extracted of more than 16000 scientific articles, almost 70% of these devices have a n-i-p architecture and active area smaller than 1 cm^2 (see **Figure 1a**). This highlighted a big research activity in small devices, contrary to devices with areas closer to commercial applications. In fact, in the case of minimodule (cells connected in series and A > = 10 cm^2), there are only 36 devices reported. Interestingly, at this scale the p-i-n architecture account for almost 42% of the devices reported. Thus, this cell configuration is of high interest for scaling up the perovskite technology since it has demonstrated higher stability, reproducibility and can be made from materials of lower cost such Nickel oxide (NiOx) as hole transport layer instead of spiro commonly used for N-I-P devices [1].

On the other hand, related to the reported stability measurements, the T_{80} which is the time at the device achieves 80% of its initial Power Conversion Efficiency (PCE), is plotted against measurement time (**Figure 1b**). Here, the **gray** area in the plot comprises 92.5% of the PSCs which have been tested for periods lower than 1600 hours, far from the temporal stability offered by commercial technologies. A deeper analysis of the data reveals that only 37 devices (~0.09%) have been tested outdoor and only 28 devices have an area larger than 1 cm^2. The small number of

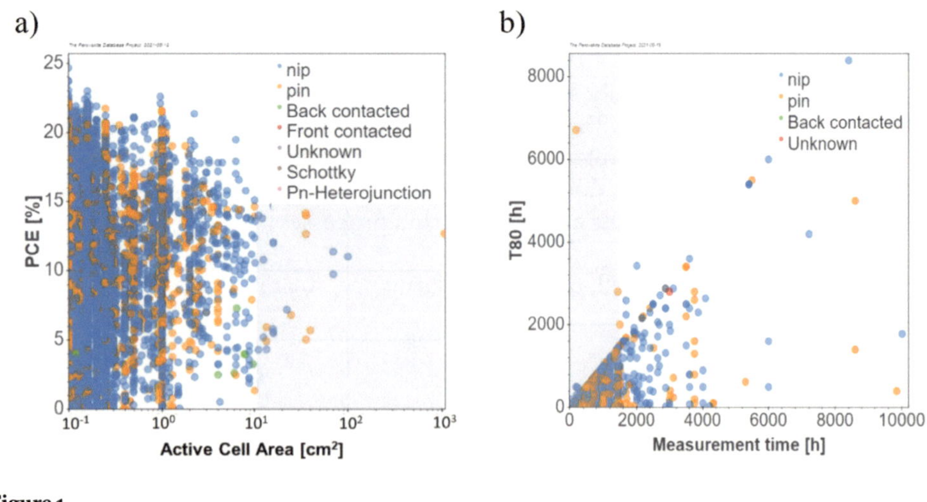

Figure 1.
State of the art of perovskite photovoltaic technology. The plots were built using the data available in the "Perovskite Database Project". (a). PCE as a function of active cell area. The gray area zone corresponds to large area devices (>10 cm^2). Colors correspond to the cell architecture. (b) T_{80} as a function of measurement time. The gray area corresponds to devices measured for less than 1500 hours. Colors correspond to the cell architecture.

publications dealing with large area PSCs tested under real conditions shows a void in the scientific literature which need to be overcome for commercialization of perovskite technology, considering that the processing techniques and materials used for the fabrication of this PSC are hardly applicable in large area devices, and the last PCE record for PSC of 25.2% is close to commercial technologies such as silicon.

Altogether, this analysis of scientific literature reveals the urgent necessity of studies dealing with performance measurements of large devices operated outdoors, tested for longer periods, and produced by solution-based methods compatible with scalable fabrication. In this chapter, we report a compilation of several studies developed in our group which report the fabrication and electrical performance characterization of perovskite minimodules tested in outdoor conditions for periods as long as 2500 hours. The p-i-n mesoporous structure of the minimodules allows fabrication of devices on substrates of areas up to 100 cm^2 and showed improved stability and reproducibility of the perovskite films on large-areas [1]. Moreover, the volatile solvent methylamine-acetonitrile used for the perovskite precursor its compatible with scalable fabrication methods [2]. To measure outdoor performance, we propose a methodology based on the international standard IEC 61853–1 [3]. From the temporal evolution of the maximum power (Pmax) of solar minimodules tracked at the Solar Cell Outdoor Performance Laboratory (OPSUA) of the University of Antioquia, we have collected statistical data which allowed the measurement of lifetime (T_{80}) as well as the temporal evolution of the ideality factor (n_{id}). From these parameters, it is possible to extract the degradation rate and monitor the evolution of the degradation processes. Moreover, the information collected enable tracking the physical processes occurring in the device under real conditions. Finally, the linear relationship between the time at which the module reaches $n_{ID} = 2$ (Tn_{ID2}) and T_{80}, suggested the complementarity of these two parameters. This complementarity has important implications for improving the characterization and understanding of the degradation processes and, consequently, for the PSC's outdoor optimization.

2. Photovoltaic performance

The performance of photovoltaic devices is conventionally characterized by the I-V curve according to temperature and illumination conditions. Accordingly, the Standard Test Conditions (STC) defined as 1000 W/m^2 of irradiance and 25°C of cell temperature is the most common conditions to measure the I-V curve in order to extract the main parameters such as open circuit voltage (V_{oc}), short circuit current (I_{sc}), maximum power (P_{max}), and estimate the efficiency [4]. STC is commonly measured in indoor, addressed in the datasheet, and used to compare the progress of solar technologies [5].

Related to outdoor performance, International Electrotechnical Commission (IEC) published a series of standards in IEC 61853 intended to establish the requirements for evaluating the performance of all photovoltaic technologies in term of power [6] or term of energy and performance ratio (IEC 61853–3). In the case of energy, it is worth noting that the study involves the inverter performance. For that, the performance depends on the evaluated solar device, the electronic device, and the maximum power tracking algorithm [7]. In the case of power, the study is mainly based on the parameters extracted for the I-V, in this regard, the standard called "Photovoltaic (PV) module performance testing and energy rating – Part 1: Irradiance and temperature performance measurements and power rating" allows the validation of the devices status of any solar technology defining a pass/fail criteria, in which the success is reached if the power rating conditions (PRC) measured fall within the power range specified by the manufacturer [6]. Corresponding the conditions for the five power rating conditions to:

- Standard Test Condition (STC), defined at 1000 W/m^2 and cell temperature of 25°C
- Nominal Operating Cell Temperature (NOCT), defined at 800 W/m^2 and ambient temperature of 20°C
- Low Irradiance Condition (LIC), defined at 200 W/m^2 and cell temperature of 25°C
- High Temperature Condition (HTC), defined at 1000 W/m^2 and cell temperature of 75°C
- Low Temperature Condition (LTC), defined at 500 W/m^2 and cell temperature of 15°C)

Consequently, the evaluation of the standard IEC61853–1 according to the five PRC provides a complete device characterization under various values of irradiance and temperature, observing the impact of weather variables on P_{max}, V_{oc}, I_{sc}. Besides, this characterization could be carried out using a solar simulator and specialized equipment to set up the temperature and illumination conditions or in natural sunlight with and without a tracker.

In the case of emerging technologies such as PSCs, no international standards have been fully established and adopted, and most published works have focused on laboratory-scale cells to evaluate the performance, stability and degradation of this technology [8–14]. Thus, because this technology is in its infancy, insufficient data are available to fully establish or identify the degradation modes and mechanisms of PSCs, the impact on outdoor performance evolution including large devices [15, 16].

Related to outdoor performance evaluation of perovskite, the International Summit on Organic Photovoltaic Stability (ISOS) suggests 3 protocols [17]. ISOS-O-1 suggests periodically record the J-V curve at STC using a solar simulator (indoor). ISOS-O-2 suggests periodically record the J-V curve under outdoor conditions, keeping the device at the Maximum Power Point (MPP) or open-circuit voltage. In contrast, the ISOS-O-3 suggests periodically recording the J-V curve using a solar simulator, keeping the device at MPP under outdoor conditions. Despite that these cases are related to outdoor, the electrical characterization is usually carried out under indoor conditions to evaluate the performance degradation or device stability, this obviously gives most of the time to over or sub estimate the real device behavior. On the other hand, the protocols suggest reporting the normalized data (per unit) considering the first value to normalize the data and obtain the degradation behavior to estimate degradation rate or lifetime. In this regard, a broadly-supported consensus statement on reporting data related to stability assessment was published, highlighting certain particularities of PSC technology that must be taken into account [14].

3. Outdoor performance

To evaluate the outdoor performance of solar devices under natural sunlight without a tracker, a Solar Cell Outdoor Performance Laboratory of the University of Antioquia (OPSUA, **Figure 2**) was implemented in Medellín-Colombia (6° 15′ 38″

DOI: http://dx.doi.org/10.5772/intechopen.100437

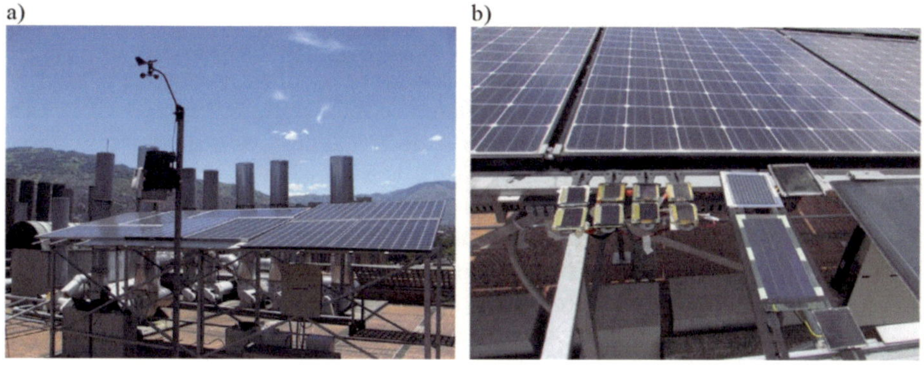

Figure 2.
Solar Cell Outdoor Performance Laboratory of the University of Antioquia (OPSUA). a) commercial technologies. b) perovskite devices.

N 75° 34′ 05″W, facing south at a fixed angle of 10°). In this laboratory, the performance of photovoltaic technologies could be evaluated in terms of power measuring the I-V curve of devices under different illumination and temperature conditions according to IEC 61853–1.

To measure the I-V curve of solar devices different hardware techniques have been used [18], intended to improve portability, local data storage, low cost, and faster response during measurement. Two different types of I-V curve tracers (solar analyzer, SA) have been recently developed [19]. One SA is based on the capacitive load technique due to simplicity, power dissipation, and cost concerns. This prototype is intended for devices that operate for voltages up to 250 V and currents up to 12.5 A (solar panels, **Figure 3a**), measuring the I-V curve from short circuit to open circuit. The other SA is based on the four-quadrant DC supply due to greater flexibility on sweep direction and speed, allowing a complete characterization including the reverse bias region and voltages higher than the open-circuit voltage, making it practical for hysteresis measurements. This prototype is intended for devices that operate for voltages up to 8 V and currents up to 3 A (solar cells and solar minimodules, **Figure 3b**). Moreover, to optimize the I-V curve process and store records, an embedded computer (Raspberry Pi) is connected to control each SA and measure the I-V curve every minute.

3.1 Methodology to estimate the performance from outdoor tests

The monitoring system implemented in OPSUA, allows to record, and store the weather variables (irradiance and temperatures), and electrical data (I-V curve). These data are uploaded to the remote server to provide backup, and centralize the

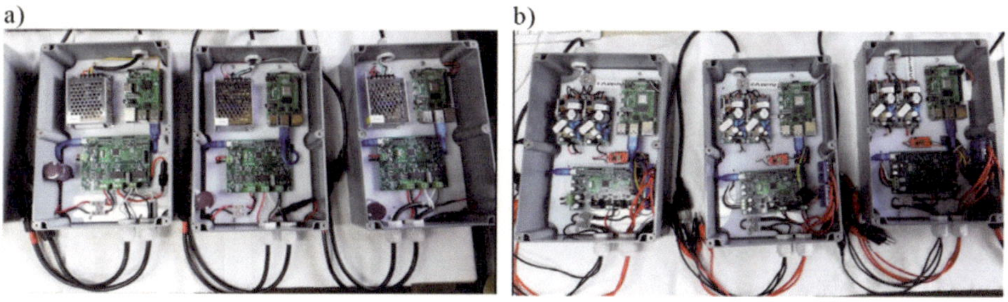

Figure 3.
Solar analyzers (SA) to tracer the I-V curve. a) SA for solar panels and b) SA for solar cells or minimodules.

process to estimate the Outdoor performance according to the flowchart shows in **Figure 2** considering the power rating conditions suggested by IEC 61853–1. In this sense, the implemented procedure to estimate the outdoor performance could be divided into two main parts: one part is related to the databases of exposed devices that are created by merging the weather and electrical data, another part is related to the analysis, **Figure 4**.

In brief, from the I-V curves, photovoltaic parameters such as V_{oc}, the short-circuit current (I_{sc}), the fill factor (FF), the photoconversion efficiency, and P_{max} were extracted. The irradiances and ambient temperatures were also recorded during the I-V measurement (synchronously). Subsequently, the raw data are filtered based on the linearity determination criterion to minimize transient effects related to changes in irradiance, shadowing caused by clouds or droplets, or atypical data [20]. Hence, the best-fit data with a deviation of ±5% were selected as the filtered data. These filtered data are used to perform the analyses:

a. Outdoor performance (OP): the impact of weather variables on performance is estimated and depicted in form of map to illustrate the performance as a function of irradiance and temperature, representing the average outdoor performance during the evaluated time and indicating the operative range of the involved variables (**Figure 5a**). A full description of these maps was shown in previous work [19–21].

b. Power rating conditions (PRCs): the data shown in the form of maps (the average OP) are filtered considering a deviation of 5% from the irradiance levels corresponding to the power rating conditions indicated by IEC 61853–1, that is, 1000 W/m^2, corresponding to standard test conditions (STC); 800 W/m^2, corresponding to NOCT conditions; 500 W/m^2, corresponding to low-temperature conditions (LTC); and 200 W/m^2, corresponding to low-irradiance conditions (LIC). Hence, the data deviation of each power rating conditions is considered by the median value of the data considered in each sampling time, **Figure 5b**.

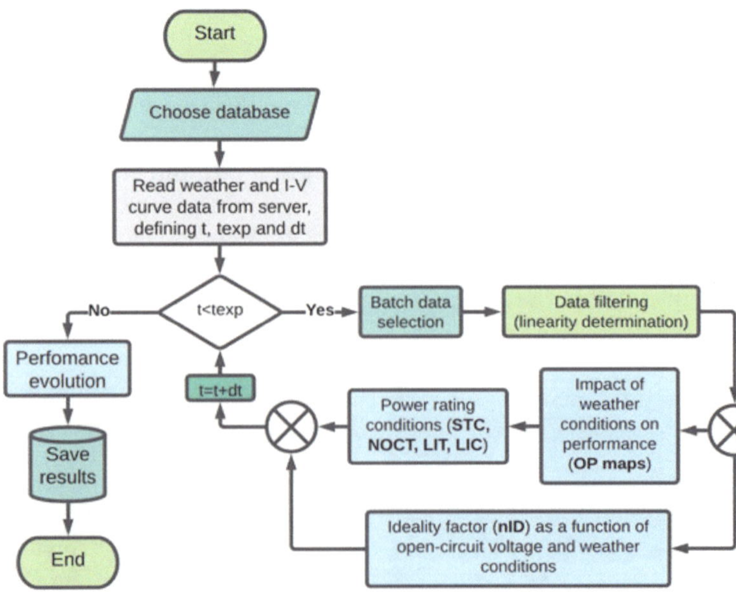

Figure 4.
Flow chart to obtain the outdoor performance. Procedure to analyze and process the databases to obtain the impact of weather variables on outdoor performance (OP), power rating conditions (PRCs), and ideality factor (nID).

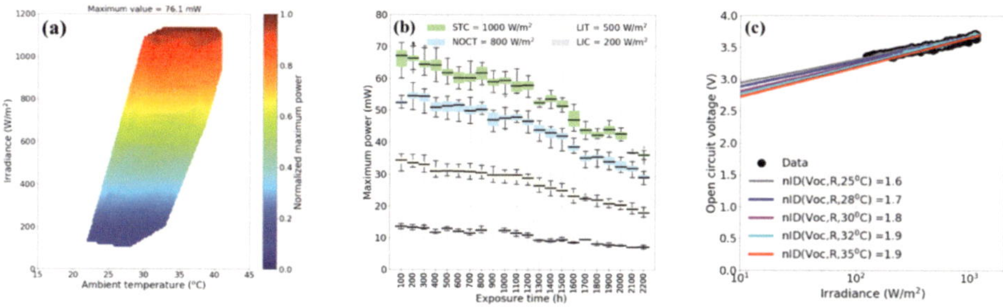

Figure 5.
Outdoor analysis. a) Outdoor performance displayed in the form of a map to show the impact of weather variables on performance. b) Calculated PRCs over time by filtering the data by batches considering the irradiance levels defined by the power rating conditions. c) Ideality factor estimated considering the ambient temperature and the data shown in figure a).

c. Ideality factor (n_{ID}): n_{ID} is calculated following the procedure shown [21], considering the filtered data and processing the data at different ambient temperatures between 25 and 35°C (**Figure 5c**). Consequently, because different ideality factors can be calculated, the median value is considered the average value to consider the data dispersion representing the evaluated time window.

Finally, it is worth noting that these analyses are considered by the loop. For that, it is possible to divide the data by sets (batches) to estimate the average impact of weather variables on performance, the power rating conditions and ideality factor in every dataset (batch). Therefore, it is possible to analyze the performance evolution or power loss tendency (degradation-shape).

3.2 Perovskite outdoor performance

In OPSUA, perovskite minimodules (PMM) with inverted structure, large areas and different cells in series have been tested [1, 20, 21]. **Figure 6** shows the performance for three representative PMMs during the first 100 h of exposure in the form of contour. These maps represent the average performance allowing observation of the impact of weather variables on P_{max}, V_{oc}, and I_{sc}. In this cases is noting that the performance for the samples show in **Figure 6a, b, d, e, g** and **h**, follow similar and expected behavior for P_{max}, I_{sc}, and V_{oc} according to the irradiance. However, the performance shows in **Figure 6c, f** and **i** do not exhibit this monotonic behavior, showing local maxima or minima at various irradiance levels and temperature, behavior that was correlated with fast degradation processes during the first 100 h of exposure [21]. Therefore, these maps could be used as a diagnostic tool aimed at identifying early faults or degradation processes in the devices. Moreover, the maps allow us to identify the low-temperature dependence of V_{oc}, which was identified as a competitive advantage of perovskite [20].

3.3 Perovskite degradation

Information on system state and performance collected over time is referred to as degradation data [22]. In solar devices (cells, modules, and panels), the natural indicator to evaluate the degradation is the performance, which is commonly obtained from the I-V curves data and contrasted to weather variables such as irradiance and temperature [23]. In this sense, the failure for an individual device could be defined as the time at which the output power dropped to 20% below the initial output, being this the standard definition of lifetime of photovoltaic devices (T_{80}). It depends on

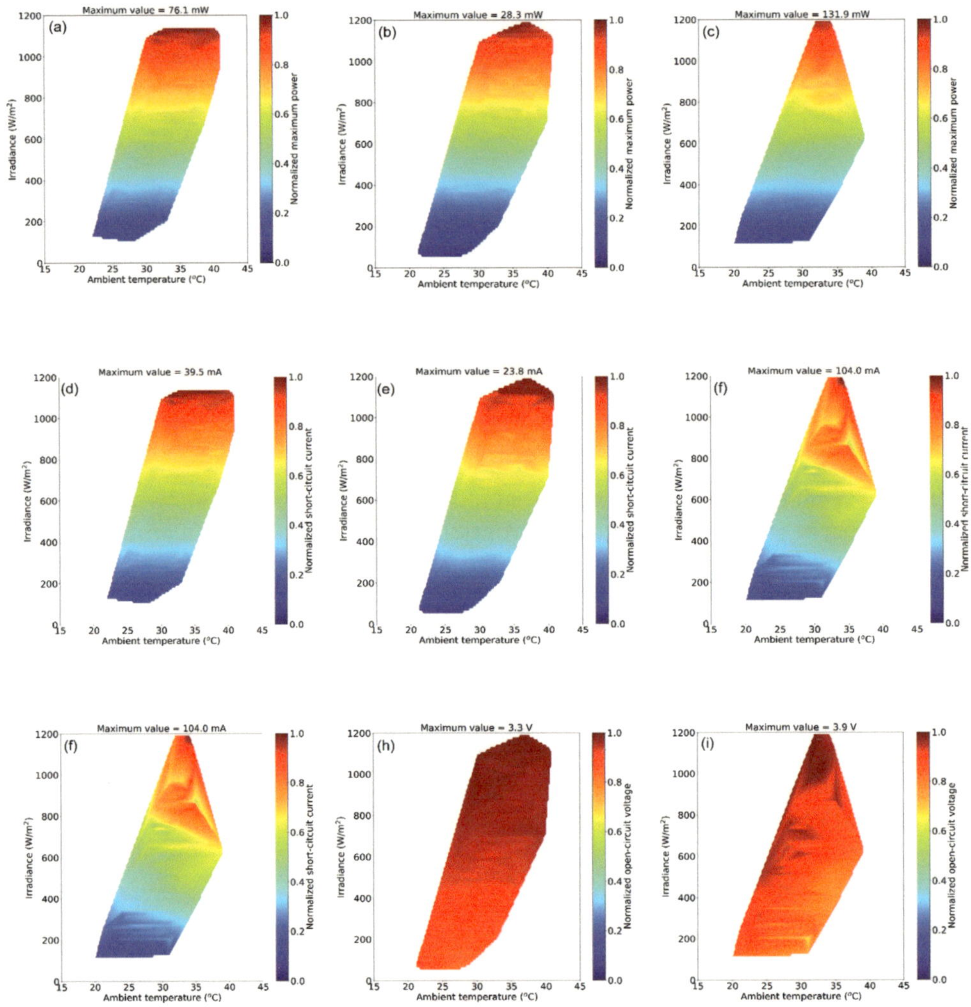

Figure 6.
Normalized outdoor performance of three representative PMM. a-c) Maximum power. d-f) Short-circuit current. g-i) Open-circuit voltage. The color bar indicates the variable range. At the top of each plot, the maximum recorded value used to normalize each variable's data is shown. These figures only considered the data during the first 100 h of exposure. Figure taken from [21]. Copyright 2021, Springer Nature.

different factors such as materials and procedures used in the device fabrication, cell interconnects, weather conditions, seasonal variations, installation conditions, shading and soiling effects, an electrical mismatch between cells, among others [24].

Currently, the long-term stability or lifetime (T_{80}) of perovskite technology on average is only a few months, even for encapsulated devices, corresponding to a degradation rate higher than 100%/year [25]. This short lifetime is the result of intrinsic and external aspects [26]. The intrinsic degradation is mainly related to thermal and light soaking effects, while external degradation is mainly related to the ingress of water, moisture, or oxygen into the device [27, 28]. Accordingly, it is possible to find different strategies to improve the lifetime, including the encapsulation process/method, kind of perovskite (2D, 3D, Etc.), selective charge contacts or other layers involved [29], passivation of the interfaces [30] or the grain boundaries [31]. Nevertheless, because the degradation is a complex process that depends on the structure (layers) and their interfaces, the degradation studies must be conducted in the complete device and not in isolated layers [32].

In this regard, **Figure 7** shows a summary of P_{max} evolution for different minimodules evaluated under outdoor conditions in OPSUA. Here, to show the performance degradation, the evolution of the performance at NOCT was selected because this power rating conditions is defined as the equilibrium mean solar cell junction temperature of the module in the described environment [33], these conditions reflecting adequately the real operating, indicating how the module temperature is affected by solar irradiation, ambient temperature and thermal properties of the photovoltaic material [34]. In addition, these conditions are included on the datasheets for commercial solar modules, and it can be reached outdoors [19] whiles the STC conditions rarely occur outdoor [35].

The evaluated outdoor samples exhibited three different P_{max} evolution patterns over time, named convex, linear, and concave patterns, because of the shapes they exhibit. These three distinctive patterns are commonly described for degradation processes in the literature to study possible degradation paths and estimate the failure time [22]. Therefore, these P_{max} behaviors were fitting to linear models to estimate the degradation rate and T_{80}. Thus, the T_{80} values for the samples were analyzed in **Figure 7b** in the form of probability distribution plot. This plot indicates that the double Weibull distribution is more suitable to represent the data distribution with a shape parameter of c = 0.6, suggesting higher early mortality of devices and that the failure rate decreases over time. Also, this result suggested 3 degradation levels based on the likelihood of T_{80} values. The first was related to the most likely and faster degradation ($T_{80} < 200$ h), the second related to moderate degradation ($350 < T_{80} < 700$ h), and the third related to lower degradation ($T_{80} > 1250$ h). These levels or groups agree with the degradation shape of concave, linear, and convex patterns, respectively. Therefore, all the analyzed minimodules present behavior that can be statistically associated with these three patterns. Finally, **Figure 7c** highlights the behavior of one representative sample of each pattern, corresponding to the lower degradation rate of the initial P_{max} of 0.29%/day to the convex pattern, a moderate degradation rate of 1.39%/day to the linear pattern, and faster degradation rate of 7.68%/day to the concave pattern.

It is worth noting that these three P_{max} patterns have all been observed for PSCs [25]. For instance, the convex pattern has been observed for encapsulated PSCs stored at room temperature [36]. The linear and convex patterns have been observed in nonencapsulated cells under controlled relative humidity conditions [37],

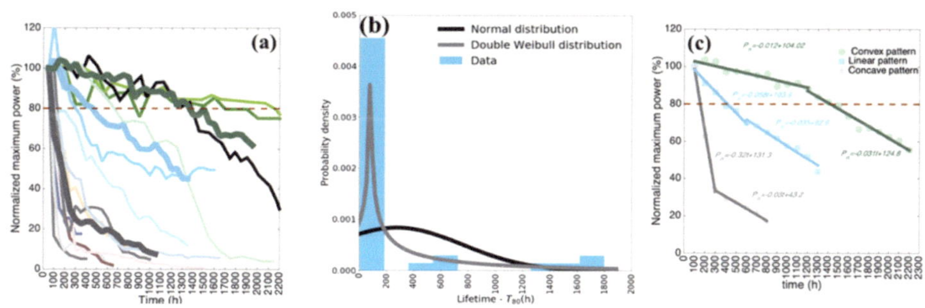

Figure 7.
Summary of P_{max} evolution at NOCT from the outdoor test. a) representative power loss tendencies for perovskite minimodules. b) Probability density of T_{80} for evaluated samples. Corresponding the blue bars to T_{80} data, black line to data fit to Normal Distribution function (being the parameters: μ = 276.8 and std. = 478) and green line to data fit to Double Weibull Distribution function (being the parameters: c = 0.6, loc = 80 and scale = 453.7). c) Degradation Rate (DR) of three representative patterns of each group according to degradation levels suggested by the probability function distribution of T_{80}. Corresponding the DR to the slope of linear fitting showed (color lines). The slope of the fitted curves provides the degradation rate in %/h units.

while the concave pattern has been observed in encapsulated cells exposed to different levels of sunlight [9]. In addition, this shape has also been observed in nonencapsulated PSCs under higher relative humidity [11], and in nonencapsulated devices tested under air exposure [38]. Nevertheless, in the case of controlled atmospheres is possible to correlate the performance behavior with the physical origins of the degradation, whereas under outdoor conditions is not easy to determine the origin of the degradation.

In the case of the representative samples of each P_{max} pattern obtained from the outdoor tests (**Figure 7**), The degradation in P_{max} can be mainly attributed to the decreases in V_{oc} or I_{sc} for the convex and linear patterns, respectively (**Figure 8a** and **b**). In contrast, both parameters are significantly degraded for the concave pattern, as is shown in **Figure 8c**. Fitting the normalized variables concerning the initial values, to linear regression models in order to estimate the degradation rates (DR) by sections (**Figure 8d–f**), is possible to observe some features:

a. Related to P_{max} convex pattern, this variable, and the V_{oc} can be modeled by 2 linear models, corresponding the DR value for the second section to 2 times first section's value for both variables. In contrast, the I_{sc} can be modeled by only one linear model characterized by a DR value of 0.0023%/h. These results indicated that the main power drop is because of the changes in V_{oc}.

b. Related to P_{max} linear pattern, this variable, and the V_{oc} can be modeled by 2 linear models. Nevertheless, the corresponding DR value for the second section for P_{max} is lower than the DR value of the first section (0.035 and 0.058%/h, respectively). For V_{oc}, the DR for the second section is

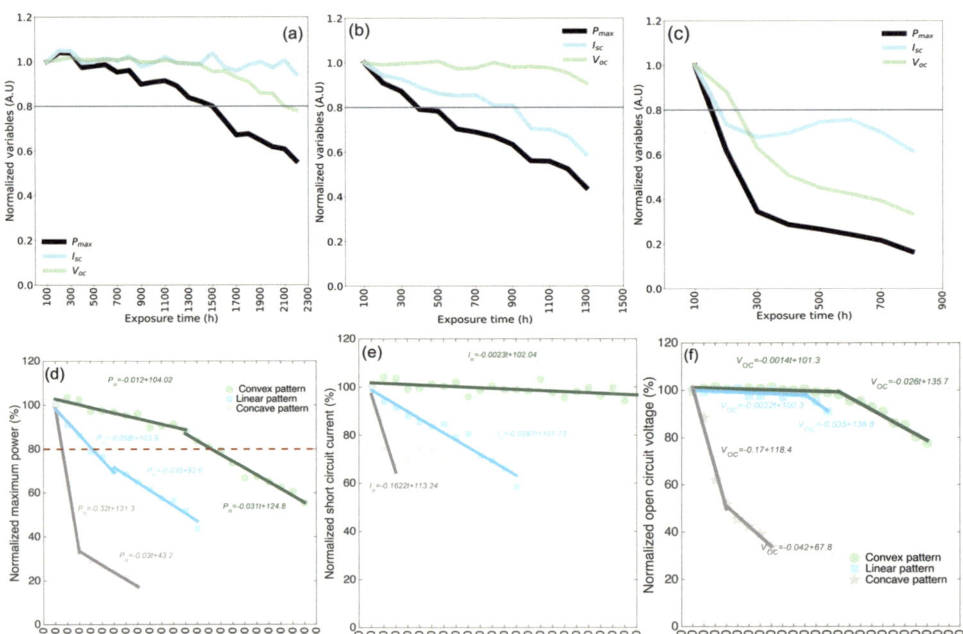

Figure 8.
Normalized behavior for photovoltaic parameters. Corresponding a) for variables related to P_{max} convex pattern, b) for P_{max} linear pattern, and c) for P_{max} concave pattern. d) Maximum power for all patterns. e) I_{sc} for all patterns and f) V_{oc} for all patterns. The green color corresponds to P_{max} convex pattern, the blue color corresponds to P_{max} linear pattern, and the gray color corresponds to P_{max} concave pattern. Solid lines correspond to the fits and color markets to the data.

16 times the first section's value (0.035 and 0.0022%/h, respectively). I_{sc} can be modeled by only one linear model characterized by a DR value of 0.0029%/h. These results indicated that the main power drop is because of the changes in I_{sc}.

c. Related to P_{max} concave pattern, all variables can be modeled by 2 linear models, corresponding the second section's DR values to the lower DR values for each variable. Moreover, in the first section, similar DR values for V_{oc} and I_{sc} were obtained. These results indicated that the power drop is because of both variables' changes (I_{sc} and V_{oc}).

d. Accordingly, P_{max} convex and linear patterns share similar trends, characterized by constant DR for I_{sc}, indicating a constant reduction in the charge extraction. Also, both V_{oc} patterns shown a double increase in the DR after a specific time (1400 and 1000 h for convex and linear patterns). This fact could be associated with surface traps allowing the charge recombination, increasing the device's series resistance, and causing the voltage drop.

3.4 Ideality factor (n_{ID})

The ideality factor also called the quality factor or shape curve factor, is one of the most reported parameters for different solar cell technologies. This parameter has been used to define the electrical behavior of solar devices due to its relationship with conduction, transport, recombination, and behavior at interface junctions, providing direct information on the dominating recombination processes. For silicon, n_{ID} has been widely studied, reporting values between 1 and 2. Close to 1 indicates ideal junctions, equal to 2 is correlated with the degradation of the solar cell, non-uniformities on recombination centers, and shunt resistance effects [39]. This parameter has been estimated using the relationship between the open-circuit voltage (V_{oc}) and light intensity to overcome the effects of series resistance [40], performing numerical calculations [41], and fitting the I-V curves to equivalent circuits [42]. In the case of perovskite, a full interpretation of n_{ID} for non-encapsulated cells was reported, establishing the relationship between the dominating recombination process, light intensity and V_{oc} [43]. In addition, under dark conditions, ideality factors close to 2 have been reported due to carrier recombination and trap-assisted recombination [44, 45] and values between 1.3 and 2.5 has been estimated considering several hole transport layer thicknesses [46]. Moreover, due to the estimation of this parameter could be affected by different aspects such as hysteresis, the dependence of open-circuit voltage on light intensity and temperature, parasitic resistances, among others, the perovskite ideality factor estimated from the dark I-V curve could be higher than 2 [47]. Nevertheless, an agreement has been shown between the n_{ID} value estimated from the recombination resistance extracted through impedance/frequency-response (IFR) analysis and the value calculated from V_{oc} at different light intensities [21, 47, 48].

On the other hand, despite that n_{ID} can be accurately estimated from the I-V curve using proper optimization methods [49], this procedure could be considered as successful if the I-V curves are well defined. In cases in which the I-V curves show an S-shape, this procedure could not be suitable, and other elements in the equivalent circuit such as another diode have been included to reproduce the I-V behavior, increasing the complexity and accurate interpretation of ideality factor [42, 50]. The S-shape behavior is associated with the chemical degradation of electrical contacts or charge accumulation on electrodes affecting the series resistance of the device [51]. Thus, to avoid the series resistances issues, another method is used to estimate the n_{ID} the relationship between V_{oc} and light intensities, Eq. (1). This equation shows the relationship between the V_{oc}, bandgap (E_g), thermal voltage, ideality factor, and logarithmic dependence on light.

$$e \cdot V_{OC} = m \cdot E_g + n_{ID} \cdot k_B \cdot T \cdot m \cdot ln \frac{G}{G_{0.}} \tag{1}$$

Where e is the electron charge, m is the number of equal cells connected in series, E_g is the light absorber bandgap, k_B is the Boltzmann constant, T is the temperature, G is the irradiance or light intensity, and G_0 is a constant with the same units than G.

Accordingly, the outdoor data related to the three representative samples correlated to each P_{max} pattern were analyzed following Eq. (1). Therefore, the average n_{ID} values were calculated from the outdoor data taking advantage of the different levels of illumination caused by day-night cycles to collect a large amount of data across a broad range of illumination conditions.

Figure 9 shows as an example the procedure to estimate the n_{ID} using the outdoor data for the sample with P_{max} convex pattern during the first 100 h of exposure. The raw data (full data during the first 100 h) are filtered, applying a linearity determination criterion to minimize the influence of atypical data or data related to unclear days, shadowing, dirt, or droplets on the surface. In this regard, the best-fit data with a deviation of ±5% are selected as the filtered data, representing the average conditions during the exposure time, **Figure 9a**. Then, the data are filtered by the ambient temperature to extract the V_{oc} and irradiance values correlated to this temperature. These V_{oc} and irradiance values are used in accordance with Eq. (1) to estimate the n_{ID} values, **Figure 9b**. Accordingly, to consider the deviations from the average value due to temperature changes, n_{ID} values are estimated by calculating this parameter in the measurement sets, filtering the data by temperatures between 25 and 37°C in steps of 2°C with a deviation of ± 1°C.

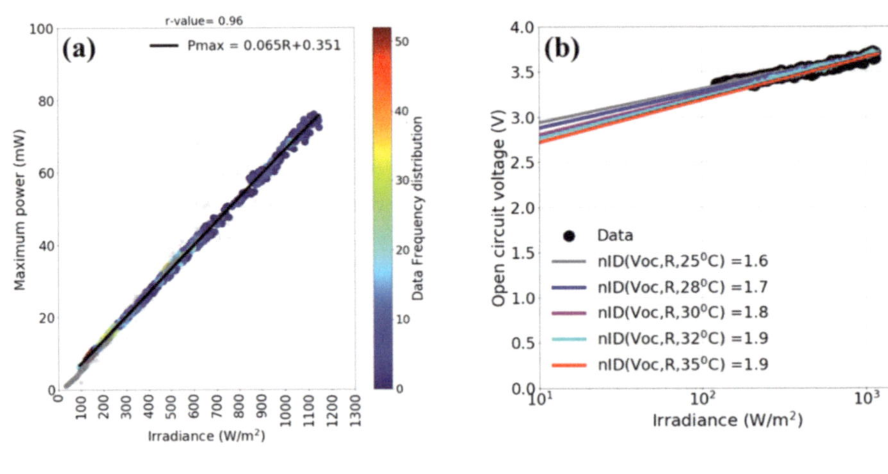

Figure 9.
*Procedure to estimate n_{ID} from Outdoor Data. a) Linearity determination criteria to filter the data from atypical data. At the top is shown the coefficient of determination (r-value), estimated from the fitting process between P_{max} and irradiance. The gray points correspond to data measured in the considered exposure time, and the color dots correspond to data considered as filtered and representative of the dataset. The filtered data correspond to the average conditions during the 100 h of exposure. b) Ideality factor estimated considering different ambient temperatures and the data registered between 0 and 100 h of exposure. The parameters estimation of n_{ID} for the linear regression in accordance with Eq. (1) is shown in **Table 1**.*

Temperature (°C)	n_{ID}	r-value	Slope	Intercept
25	1.58	0.98	0.16	2.57
28	1.67	0.98	0.17	2.48
30	1.81	0.99	0.18	2.37
32	1.86	0.99	0.19	2.32
35	1.91	0.97	0.20	2.25

Table 1.
Parameters related to the fit of V_{oc} and Irradiance to estimate n_{ID}.

Then the mean value of each dataset is considered as the average n_{ID} value. The results during the exposure time for the representative samples are shown in **Figure 10** in the form of boxplots to illustrate the deviations from the average value and data distribution for the samples analyzed.

The n_{ID} results showed in **Figure 10** also exhibits three distinct evolution patterns or shapes. Precisely, a convex P_{max} evolution pattern corresponds to a concave n_{ID} evolution pattern (**Figure 10a**) and *vice versa* (**Figure 10c**). In contrast, for a linear P_{max} pattern, a linear n_{ID} pattern is observed (**Figure 10b**). In the cases of convex P_{max}/concave n_{ID} patterns (**Figures 9** and **10a**) and linear patterns (**Figures 9** and **10b**), at times earlier than T_{80} (1442.2 and 414.2 h, respectively), n_{ID} takes values between 1 and 2, indicating bulk Shockley-Read-Hall (SRH) recombination [43]. For longer times, n_{ID} exhibits values above 2, characteristic of a multiple-trap distribution, originating from the formation of trap states, causing the performance degradation, as pointed out by Khadka et al. [36]. In concave P_{max}/convex n_{ID} patterns (**Figures 9** and **11c**), the initial values of n_{ID} are higher than 2, indicating fast degradation in the first 100 h of exposure due to the formation of multiple trap states. Interestingly, after the initial increase in n_{ID}, a progressive decrease is observed in the concave and convex (**Figure 11a** and **c**) cases, suggesting evolution from bulk recombination to interfacial recombination. This behavior does not imply a recovery in device performance; it only indicates a transition between two different recombination regimes in the degradation process, from multiple-trap recombination to a regime with higher interfacial recombination.

It is worth noting that results in **Figure 10** were estimated in accordance with Eq. (1) using the ambient temperature to simplify the methodology and the number of variables to record in the outdoor test. Nevertheless, to determine the impact of

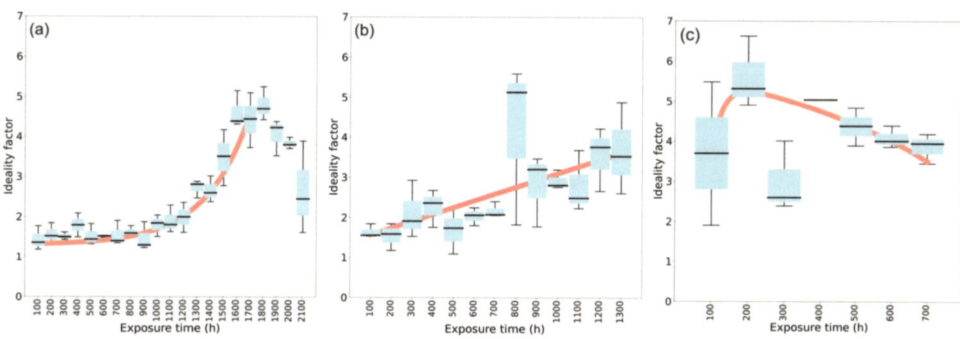

Figure 10.
Patterns of the ideality factor. a-c) Ideality factor analysis of the three representative samples: (a) n_{ID} exhibits a concave pattern, (b) n_{ID} exhibits a linear pattern, and (c) n_{ID} exhibits a convex pattern. The thick black line on each box represents the average value in the corresponding time window. The red line is included as a visual guide to illustrate the patterns. Figure taken from [21]. Copyright 2021, Springer Nature.

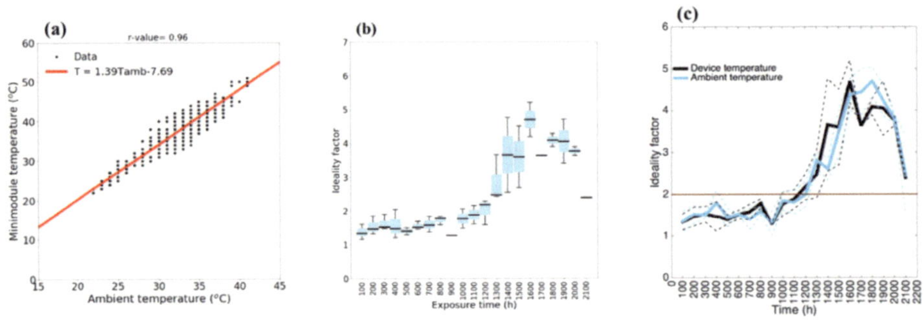

Figure 11.
Estimated ideality factor using the device temperature. a) Relationship between ambient and minimodule temperature during the first 100 h of exposure. Corresponding the points to measured data and the solid red line to the linear regression fit. The corresponding coefficient of determination of the fits is shown at the top. b) Ideality factor calculated considering the device temperature instead of the ambient temperature. The boxplots were obtained by calculating the variables observed in the measurement sets, each of which contained measurements recorded over 100 hours of outdoor exposure. The thick black line on each box represents the median value, while the blue boxes represent 50% of the data in the corresponding time window. c) Comparison of ideality factor calculated using the device and ambient temperature. Solid lines of each color correspond to the estimated average n_{ID} values and dotted lines to the standard deviation of the variables. These values were estimated by calculating the n_{ID} values in the corresponding time window for ambient temperatures ranging between 25 and 35°C and device temperature between 25 and 45°C in steps of 2°C.

the device temperature in the estimation of the ideality factor, the data related to P_{max} convex pattern were used to estimate this parameter. In this sense, **Figure 11a** shows the relationship between ambient and device temperature, which depicted a linear behavior characterized by the figure's inset equation and demonstrated by the r-value close to 1 shown at the top. **Figure 11b** shows the results of calculating the n_{ID} using the device temperature. This behavior is very similar to the n_{ID} behavior obtained in **Figure 10a** using the ambient temperature. Therefore, to compare in a better way both behaviors, **Figure 11c** shows the average values and standard deviation of the variable in the measurements sets, which contain the data over 100 h of exposure. Hence, the error between both estimations was 4.3%, which was calculated using the Euclidean norm between the difference of both estimations. Consequently, it was possible to validate that using the ambient temperature instead of device temperature allowed us to estimate reliable values to track the evolution (because this estimation introduces a relative error lower than 5%) and simplifies the number of outdoor variables.

4. Conclusions

Outdoor performance of emerging halide perovskite photovoltaics is just starting to become an important issue to determine the technology behavior under more realistic conditions. Specific standards are required to be applied for particular technologies specially for halide perovskite solar devices, as their behavior not necessarily follows the typical silicon or other traditional semiconductors based technology. However, general standards like IEC 61853–1 allows to stablish valid criteria to study these emerging devices. Parameters as lifetime (T_{80}) and ideality factor (n_{ID}) can be estimated from the maximum power output trends and open-circuit voltage (V_{oc}) dependence on irradiance and ambient temperature from outdoor data. Moreover, the values can provide physical insight into the recombination mechanism dominating the performance during the exposure. Perovskites degradation indicates both bulk Shockley-Read-Hall (SRH) recombination and multiple-trap states for higher values of ideality factor. These methodologies allow to continue studying different architectures, composition, encapsulating methods to improve the performance of the perovskite photovoltaics under outdoor conditions before they can be release into market.

Acknowledgements

The authors gratefully acknowledge the financial support provided by the Colombia Scientific Program within the framework of the call *Ecosistema Científico* (contract no. FP44842—218-2018). Esteban Velilla thanks to the Ministry of Science, Technology and Innovation (Minciencias) for national doctoral scholarship number 727–2015 (contract no. FP44842–124-2017). Iván Mora-Seró thanks the European Research Council (ERC) funding for this work via a Consolidator grant (no. 724424—No-LIMIT).

Conflict of interest

The authors declare no conflict of interest

Author details

Esteban Velilla Hernández[1,2]*, Juan Bernardo Cano Quintero[1], Juan Felipe Montoya[2], Iván Mora-Seró[3] and Franklin Jaramillo Isaza[2]

1 Grupo en Manejo Eficiente de la Energía (GIMEL), Universidad de Antioquia, UdeA, Medellín, Colombia

2 Centro de Investigación, Innovación y Desarrollo de Materiales (CIDEMAT), Universidad de Antioquia, UdeA, Medellín, Colombia

3 Institute of Advanced Materials (INAM), Jaume I University, Castelló de la Plana, Spain

*Address all correspondence to: esteban.velilla@udea.edu.co

References

[1] Ramirez D, Velilla E, Montoya JF, Jaramillo F. Mitigating scalability issues of perovskite photovoltaic technology through a p-i-n meso-superstructured solar cell architecture. Sol Energy Mater Sol Cells [Internet]. Elsevier B.V.; 2019;195:191-7. Available from: https://linkinghub.elsevier.com/retrieve/pii/S0927024819301217

[2] Wu C, Wang K, Li J, Liang Z, Li J, Li W, et al. Volatile solution: the way toward scalable fabrication of perovskite solar cells? Matter [Internet]. Elsevier Inc.; 2021;4:775-93. Available from: https://doi.org/10.1016/j.matt.2020.12.025

[3] Velilla Hernandez E. Computational Modeling of the Electrical Performance and Degradation of Third-generation Photovoltaic Modules, Under Accelerated and Real Operating Conditions [Thesis] [Internet]. Antioquia University; 2021. Available from: http://hdl.handle.net/10495/20696

[4] IEC 61215-1-2. Terrestrial photovoltaic (PV) modules – Design qualification and type approval – Part 1-2: Special requirements for testing of thin-film Cadmium Telluride (CdTe) based photovoltaic (PV) modules. 2004.

[5] Green M, Dunlop E, Hohl-Ebinger J, Yoshita M, Kopidakis N, Hao X. Solar cell efficiency tables (version 57). Prog Photovoltaics Res Appl [Internet]. John Wiley and Sons Ltd.; 2021;29:3-15. Available from: https://onlinelibrary.wiley.com/doi/10.1002/pip.3371

[6] IEC 61853-1. Photovoltaic (PV) module performance testing and energy rating - Part 1: Irradiance and temperature performance measurements and power rating. Int. Electrotech. Comm. 2011.

[7] Eltamaly AM, Farh HMH, Othman MF. A novel evaluation index for the photovoltaic maximum power point tracker techniques. Sol Energy [Internet]. Elsevier; 2018;174:940-56. Available from: https://doi.org/10.1016/j.solener.2018.09.060

[8] Holzhey P, Saliba M. A full overview of international standards assessing the long-term stability of perovskite solar cells. J Mater Chem A [Internet]. 2018;6:21794-808. Available from: http://pubs.rsc.org/en/Content/ArticleLanding/2018/TA/C8TA06950F

[9] Anoop KM, Khenkin M V., Di Giacomo F, Galagan Y, Rahmany S, Etgar L, et al. Bias-Dependent Stability of Perovskite Solar Cells Studied Using Natural and Concentrated Sunlight. Sol RRL [Internet]. Wiley; 2020;4:1900335. Available from: https://onlinelibrary.wiley.com/doi/abs/10.1002/solr.201900335

[10] Yang J, Siempelkamp BD, Liu D, Kelly TL. Investigation of $CH_3NH_3PbI_3$ Degradation Rates and Mechanisms in Controlled Humidity Environments Using in Situ Techniques. ACS Nano [Internet]. 2015;9:1955-1963. Available from: http://pubs.acs.org/doi/10.1021/nn506864k

[11] Domanski K, Alharbi EA, Hagfeldt A, Grätzel M, Tress W. Systematic investigation of the impact of operation conditions on the degradation behavior of perovskite solar cells. Nat Energy [Internet]. 2018;3:61-67. Available from: http://www.nature.com/articles/s41560-017-0060-5

[12] Cheacharoen R, Rolston N, Harwood D, Bush KA, Dauskardt RH, McGehee MD. Design and understanding of encapsulated perovskite solar cells to withstand temperature cycling. Energy Environ Sci [Internet]. Royal Society of Chemistry; 2018;11:144-150. Available from: http://dx.doi.org/10.1039/C7EE02564E

[13] Tress W, Domanski K, Carlsen B, Agarwalla A, Alharbi EA, Graetzel M, et al. Performance of perovskite solar cells under simulated temperature-illumination real-world operating conditions. Nat Energy [Internet]. 2019;4:568-74. Available from: http://www.nature.com/articles/s41560-019-0400-8

[14] Khenkin M V., Katz EA, Abate A, Bardizza G, Berry JJ, Brabec C, et al. Consensus statement for stability assessment and reporting for perovskite photovoltaics based on ISOS procedures. Nat Energy [Internet]. 2020;5:35-49. Available from: http://www.nature.com/articles/s41560-019-0529-5

[15] Hu Y, Chu Y, Wang Q, Zhang Z, Ming Y, Mei A, et al. Standardizing Perovskite Solar Modules beyond Cells [Internet]. Joule. 2019. p. 2076-2085. Available from: https://linkinghub.elsevier.com/retrieve/pii/S2542435119304180

[16] Qiu L, He S, Ono LK, Liu S, Qi Y. Scalable Fabrication of Metal Halide Perovskite Solar Cells and Modules [Internet]. ACS Energy Lett. 2019. p. 2147-67. Available from: http://pubs.acs.org/doi/10.1021/acsenergylett.9b01396

[17] Reese MO, Gevorgyan SA, Jørgensen M, Bundgaard E, Kurtz SR, Ginley DS, et al. Consensus stability testing protocols for organic photovoltaic materials and devices. Sol Energy Mater Sol Cells [Internet]. Elsevier B.V.; 2011;95:1253-67. Available from: https://linkinghub.elsevier.com/retrieve/pii/S092702481100050X

[18] Duran E, Piliougine M, Sidrach-De-Cardona M, Galan J, Andujar JM. Different methods to obtain the I-V curve of PV modules: A review. Conf Rec IEEE Photovolt Spec Conf. 2008;

[19] Velilla E, Cano JB, Jaramillo F. Monitoring system to evaluate the outdoor performance of solar devices considering the power rating conditions. Sol Energy [Internet]. Elsevier; 2019;194:79-85. Available from: https://doi.org/10.1016/j.solener.2019.10.051

[20] Velilla E, Ramirez D, Uribe J-I, Montoya JF, Jaramillo F. Outdoor performance of perovskite solar technology: Silicon comparison and competitive advantages at different irradiances. Sol Energy Mater Sol Cells [Internet]. Elsevier B.V.; 2019;191:15-20. Available from: https://doi.org/10.1016/j.solmat.2018.10.018

[21] Velilla E, Jaramillo F, Mora-Seró I. High-throughput analysis of the ideality factor to evaluate the outdoor performance of perovskite solar minimodules. Nat Energy [Internet]. Springer Science and Business Media LLC; 2021;6:54-62. Available from: http://www.nature.com/articles/s41560-020-00747-9

[22] Meeker W, Hong Y, Escobar L. Degradation Models and Analyses. Encycl Stat Sci [Internet]. Hoboken, NJ, USA: John Wiley & Sons, Inc.; 2011. p. 1-23. Available from: http://doi.wiley.com/10.1002/0471667196.ess7148

[23] IEC 61853-2. Photovoltaic (PV) module performance testing and energy rating-Part 2_ Spectral responsivity, incidence angle and module operating temperature measurements. 2016.

[24] Phinikarides A, Kindyni N, Makrides G, Georghiou GE. Review of photovoltaic degradation rate methodologies. Renew Sustain Energy Rev. [Internet]. Elsevier; 2014;40:143-152. Available from: http://dx.doi.org/10.1016/j.rser.2014.07.155

[25] He S, Qiu L, Ono LK, Qi Y. How far are we from attaining 10-year lifetime for metal halide perovskite solar cells? Mater Sci Eng R Reports [Internet]. Elsevier Ltd.; 2020;140:100545. Available from: https://linkinghub.

elsevier.com/retrieve/pii/S0927796X20300036

[26] Boyd CC, Cheacharoen R, Leijtens T, McGehee MD. Understanding Degradation Mechanisms and Improving Stability of Perovskite Photovoltaics. Chem Rev. [Internet]. American Chemical Society; 2019;119:3418-3451. Available from: http://pubs.acs.org/doi/10.1021/acs.chemrev.8b00336

[27] Bisquert J, Juarez-Perez EJ. The Causes of Degradation of Perovskite Solar Cells. J Phys Chem Lett [Internet]. American Chemical Society; 2019;10:5889-5891. Available from: https://pubs.acs.org/doi/10.1021/acs.jpclett.9b00613

[28] McLeod JA, Liu L. Prospects for Mitigating Intrinsic Organic Decomposition in Methylammonium Lead Triiodide Perovskite. J Phys Chem Lett [Internet]. 2018;9:2411-2417. Available from: https://pubs.acs.org/doi/10.1021/acs.jpclett.8b00323

[29] Heo S, Seo G, Lee Y, Seol M, Kim SH, Yun D-J, et al. Origins of High Performance and Degradation in the Mixed Perovskite Solar Cells. Adv Mater [Internet]. 2019;31:1805438. Available from: http://doi.wiley.com/10.1002/adma.201805438

[30] Yang B, Suo J, Mosconi E, Ricciarelli D, Tress W, De Angelis F, et al. Outstanding Passivation Effect by a Mixed-Salt Interlayer with Internal Interactions in Perovskite Solar Cells. ACS Energy Lett [Internet]. American Chemical Society (ACS); 2020;5:3159-67. Available from: https://pubs.acs.org/doi/10.1021/acsenergylett.0c01664

[31] Zhang S, Liu Z, Zhang W, Jiang Z, Chen W, Chen R, et al. Barrier Designs in Perovskite Solar Cells for Long-Term Stability. Adv Energy Mater [Internet]. 2020;2001610:2001610. Available from: https://onlinelibrary.wiley.com/doi/abs/10.1002/aenm.202001610

[32] Liu Z, Qiu L, Ono LK, He S, Hu Z, Jiang M, et al. A holistic approach to interface stabilization for efficient perovskite solar modules with over 2,000-hour operational stability. Nat Energy [Internet]. Springer US; 2020; Available from: http://dx.doi.org/10.1038/s41560-020-0653-2

[33] IEC 61215-1-4. IEC 61215-1-4. Terrestrial photovoltaic (PV) modules – Design qualification and type approval – Part 1-4: Special requirements for testing of thin-film Cu(In,Ga)(S,Se)2 based photovoltaic (PV) modules. 2016;

[34] Makrides G, Zinsser B, Norton M, E. G. Performance of Photovoltaics Under Actual Operating Conditions. Third Gener Photovoltaics [Internet]. InTech; 2012. Available from: www.intechopen.com

[35] Dash PK, Gupta NC, Rawat R, Pant PC. A novel climate classification criterion based on the performance of solar photovoltaic technologies. Sol Energy [Internet]. Elsevier Ltd.; 2017;144:392-8. Available from: http://dx.doi.org/10.1016/j.solener.2017.01.046

[36] Khadka DB, Shirai Y, Yanagida M, Miyano K. Degradation of encapsulated perovskite solar cells driven by deep trap states and interfacial deterioration. J Mater Chem C [Internet]. The Royal Society of Chemistry; 2018;6:162-170. Available from: http://xlink.rsc.org/?DOI=C7TC03733C

[37] Ma C, Shen D, Qing J, Thachoth Chandran H, Lo M-F, Lee C-S. Effects of Small Polar Molecules (MA + and H 2 O) on Degradation Processes of Perovskite Solar Cells. ACS Appl Mater Interfaces [Internet]. 2017;9:14960-14966. Available from: http://pubs.acs.org/doi/10.1021/acsami.7b01348

[38] Lee H, Lee C, Song H-J. Influence of Electrical Traps on the Current Density Degradation of Inverted Perovskite Solar Cells. Materials (Basel) [Internet].

MDPI AG; 2019;12:1644. Available from: https://www.mdpi.com/1996-1944/12/10/1644

[39] Jain A, Kapoor A. A new method to determine the diode ideality factor of real solar cell using Lambert W-function. Sol Energy Mater Sol Cells. 2005;85:391-396.

[40] Kerr MJ, Cuevas A. Generalized analysis of the illumination intensity vs. open-circuit voltage of solar cells. Sol Energy. 2004;76:263-267.

[41] Santakrus Singh N, Jain A, Kapoor A. Determination of the solar cell junction ideality factor using special trans function theory (STFT). Sol Energy Mater Sol Cells [Internet]. Elsevier; 2009;93:1423-1426. Available from: http://dx.doi.org/10.1016/j.solmat.2009.03.013

[42] Bashahu M, Nkundabakura P. Review and tests of methods for the determination of the solar cell junction ideality factors. Sol Energy [Internet]. 2007;81:856-863. Available from: https://linkinghub.elsevier.com/retrieve/pii/S0038092X06002842

[43] Tress W, Yavari M, Domanski K, Yadav P, Niesen B, Correa Baena JP, et al. Interpretation and evolution of open-circuit voltage, recombination, ideality factor and subgap defect states during reversible light-soaking and irreversible degradation of perovskite solar cells. Energy Environ Sci [Internet]. Royal Society of Chemistry; 2018;11:151-65. Available from: http://dx.doi.org/10.1039/C7EE02415K

[44] Agarwal S, Seetharaman M, Kumawat NK, Subbiah AS, Sarkar SK, Kabra D, et al. On the Uniqueness of Ideality Factor and Voltage Exponent of Perovskite-Based Solar Cells. J Phys Chem Lett [Internet]. 2014;5:4115-21. Available from: https://pubs.acs.org/doi/10.1021/jz5021636

[45] Wetzelaer G-JAH, Scheepers M, Sempere AM, Momblona C, Ávila J, Bolink HJ. Trap-Assisted Non-Radiative Recombination in Organic–Inorganic Perovskite Solar Cells. Adv Mater [Internet]. 2015;27:1837-41. Available from: http://doi.wiley.com/10.1002/adma.201405372

[46] Marinova N, Tress W, Humphry-Baker R, Dar MI, Bojinov V, Zakeeruddin SM, et al. Light Harvesting and Charge Recombination in CH 3 NH 3 PbI 3 Perovskite Solar Cells Studied by Hole Transport Layer Thickness Variation. ACS Nano [Internet]. 2015;9:4200-9. Available from: http://pubs.acs.org/doi/abs/10.1021/acsnano.5b00447

[47] Almora O, Cho KT, Aghazada S, Zimmermann I, Matt GJ, Brabec CJ, et al. Discerning recombination mechanisms and ideality factors through impedance analysis of high-efficiency perovskite solar cells. Nano Energy [Internet]. Elsevier Ltd.; 2018;48:63-72. Available from: https://doi.org/10.1016/j.nanoen.2018.03.042

[48] Contreras-Bernal L, Ramos-Terrón S, Riquelme A, Boix PP, Idígoras J, Mora-Seró I, et al. Impedance analysis of perovskite solar cells: A case study. J Mater Chem A [Internet]. Royal Society of Chemistry; 2019;7:12191-200. Available from: http://xlink.rsc.org/?DOI=C9TA02808K

[49] Velilla E, Cano J, Jimenez K, Valencia J, Ramirez D, Jaramillo F. Numerical Analysis to Determine Reliable One-Diode Model Parameters for Perovskite Solar Cells. Energies [Internet]. 2018;11:1963. Available from: www.mdpi.com/journal/energies

[50] Cotfas DT, Cotfas PA, Kaplanis S. Methods to determine the dc parameters of solar cells: A critical review. Renew Sustain Energy Rev. [Internet]. Elsevier; 2013;28:588-596. Available from: http://dx.doi.org/10.1016/j.rser.2013.08.017

[51] Guerrero A, You J, Aranda C, Kang YS, Garcia-Belmonte G, Zhou H, et al. Interfacial Degradation of Planar Lead Halide Perovskite Solar Cells. ACS Nano [Internet]. 2016;10:218-224. Available from: https://pubs.acs.org/doi/10.1021/acsnano.5b03687

Chapter 6

Hybrid Perovskite Thin Film

Akin Olaleru, Eric Maluta, Joseph Kirui and Olasoji Adekoya

Abstract

Hybrid perovskite thin film offers diverse advantages like low cost deposition techniques, less material consumption and superior optoelectronic properties. These merits including high voltage and high efficiency performance in a wide range of high light intensity are sufficient to distinguish perovskite thin films/devices from their contenders as a thin film technology with greater potential for industrial applications. Perovskite thin film technology demonstrates potency in a variety of applications in optoelectronic devices especially photovoltaic applications. The National Renewable Energy Laboratory (NREL) of the USA categorizes a number of thin films technologies including perovskite thin film, as emerging photovoltaics with the bulk of them yet to be commercially applied but are still in the research or developmental stage. In this chapter, various processing methods and material combinations as well as current trends in this technology are subjects of discussion.

Keywords: Perovskite thin film, deposition techniques, low-cost, properties, solar energy

1. Introduction

Thin film is a two dimensional material layer ranging from fractions of a nanometer (single layer) to various micrometers in thickness deposited on a substrate, including plastic, glass or metal, so as to attain properties that cannot be realized at all with the same material in its bulk form. Thin-film is developed by depositing a single or multiple layers of absorbing materials on substrate for photovoltaic and optoelectronic applications. The definite property of a thin film is due to the special approach of creating it, by means of evolutionary addition of atoms or molecules. Thickness is the basic property of thin film and is firmly connected to diverse properties which vary with thickness.

1.1 Different types of thin film devices?

There are various classes of thin film devices which are widely employed in optical devices and semiconductors as well as other applications. Thin film devices are used in the form of optical filters, such as photograph, microscopes. Thin film semiconductor also finds applications in fields of microelectronics, micro-circuitry, solar cells and integrated circuits are possible type of thin film devices. In addition, common thin film devices are mirrors which are made of a thin film of reflective material like aluminum which is attached to the glass substrate. Presently in the market, thin films are being introduced to deliver drug. The film permits the drug to penetrate the bloodstream rapidly in complete contrast swallowed pills or liquid suspensions.

In this chapter, we are going to narrow down to thin film photovoltaics (PVs) major. The reasons for wide usage are not far-fetched; firstly, film thickness differs from a few nanometer (nm) to micrometer (μm) which is much thinner than the traditional crystalline silicon PV cell (c-Si) that utilizes wafers of up to 200 μm thick. Accordingly, thin film cell more flexible and lighter in weight. Secondly, thin film technologies allow drastically reduced quantities of active material in a cell, which makes the thin films much cheaper than their bulk counterparts.

Thin films photovoltaics are called based on their main light absorbing material as shown in **Figure 1**. The commercial thin film photovoltaics are mainly based on the following technologies:

- Cadmium Telluride (CdTe)
- Amorphous Silicon (a-Si)
- Copper Indium Diselenide (CIS)
- Gallium Arsenide (GaAs)

These thin film technologies are employed widely in optoelectronic devices and semiconductors, although there are different applications also. Most of the thin films used are developed out of a variation of different metals and materials, each with its own properties and applications.

Unfortunately, these earlier thin film technologies are still relatively expensive to manufacture. The cost has been a major restriction in the way of expanding the market in addition the high temperature requirements in production and the low absorption rate of photons by these films contribute to their weaknesses.

Light-harvesting layer, referred to as the main premise of any given PV technology, assumes a key function in deciding the most suitable manufacturing technique [1]. For instance, traditional silicon wafer photovoltaic cells demand energy-intensive vacuum processes and hence, normally developed in a batch process. Second generation thin-film solar cells use direct band gap micron-thick absorber (CdTe, CIGS) layer for effective conversion of solar to electric energy.

The employment of thin absorber layer eases the acceptance of high efficiency device fabrication on a variety of substrates. On the other hand, the requirement of high-vacuum and costly precursor materials restricted the commercial feasibility of thin film photovoltaic cells.

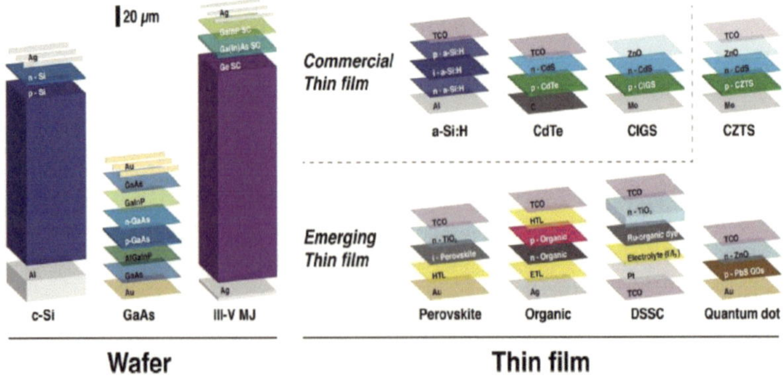

Figure 1.
PV device architectures: Wafer-based and thin-film technologies. Main absorber layers are marked by white color [1].

In recent times, the National Renewable Energy Laboratory (NREL) has categorized various thin-film technologies as emerging photovoltaics, although most of these technologies are still in the research and developmental stage, and yet to be industrially used. Many of these technologies use organic materials, like organometallic compounds and inorganic materials. These emerging thin films photovoltaics (technologies) fall under the third and fourth generations, which include:

- Organic solar cell
- A dye-sensitized solar cell (DSSC) or "Grätzel cell"
- Copper zinc tin sulfide solar cell (CZTS)
- Quantum dot solar cell
- Perovskite solar cells (PSCs)

The emerging thin films photovoltaics offered a golden opportunity for low-temperature and non-vacuum processability.

However, some of these technologies have low efficiencies, and moreover the stability of the absorbing materials is inherently too short for industrial applications. A great number of studies are ongoing on these thin film technologies as they have the potential to accomplish the goal of ushering in the much anticipated low-cost and energy-effective photovoltaic cells.

Among these emerging thin films, the perovskite thin film which serve as the focal point of this chapter provided the promising window to replace the dominant silicon photovoltaics in the market and also help mitigate the energy crisis globally. This is possible owing to desirable properties given below.

2. Materials profile for hybrid perovskite

A hybrid perovskite is a semiconductor consisting of organic, inorganic materials/species and halides. These materials are widely used as light-absorbers in perovskite solar cells. The excitement for hybrid perovskite materials has been connected to a large extent to the interesting attributes of the inorganic components, which impart thermal stability and the considerable degree of structural order [2]. In addition, the organic component properties likewise play a part in the great properties of halide perovskite including mechanical adaptability and minimal effort in production [3]. Consequently, the probability of merging the properties of normal inorganic crystals with those of organic solids has motivated recent research into the flexible properties of organic–inorganic hybrid perovskite materials.

This family of hybrid perovskite materials has been extensively employed in optoelectronics. The use of halide perovskites in PV cells has made a major quantum leap in the field of emerging PV technologies in recent years. The power conversion efficiencies (PCEs) of PSCs have demonstrated a notable rise from the pioneering value of 3.8% in 2009 to the latest record of 25.5 and 29.5% for the single junction device and the tandem with silicon (Si) respectively [4]. Besides the sphere of PV, the three-dimensional halide perovskites as well as low-dimensional perovskites have been largely investigated for applications in transistors and light emission.

The properties that render perovskite materials as the preferred choice for PV technology are summarized as follows:

2.1 Charge generation

The superb absorption properties together with low exciton binding energy enables perovskite-based solar cells to absorb light throughout visible wavelengths and beyond. This leads to a high population of charge carriers (both electrons and holes) inside the same absorbing material (ambipolar class material), which has great impact on the transport, recombination, and extraction of charges. This promotes low energy-loss by the charges produced and enhances their accumulation at the electrodes. Hence, this results in high PCEs in perovskite solar cells.

2.2 Charge transport

Unlike other PV technologies, PSCs convey both electrons and holes for extraction effectively due to its ambipolar transport properties. The superior charge-transport properties of the material with long charge carrier diffusion lengths of more than 5 μm, and corresponding lifetime of ~1 μs [5, 6] in single-crystal and poly-crystalline thin films determine the exceptional PCEs of these cells. The diffusion lengths of PSCs are greater than those of most organic solar cells prepared using solution-process technique. The diffusion lengths for both electrons and holes appear to be symmetrical in perovskite solar cells, which is essential for thin-film PV for the effective extraction of both types of carriers.

2.3 Charge recombination/energy utilization

Perovskite materials can effectively separate and transport electrons and holes as a result of quick exciton dissociation, long charge diffusion lengths and probably self-doping at the interface. Research have shown that the unusual properties of perovskites, like quick exciton dissociation and appreciable diffusion lengths, plus shallow trap states due to the presence of impurities, greatly minimize the loss of charge carrier within the perovskite bulk. The energy utilization in PSCs is very high and nearly comparable to that observed in other foremost monolithic crystalline technologies like c-Si-based solar cells.

2.4 Versatile solution processing

Perovskite materials can dissolve in organic polar solvents, thereby enabling excellent-quality thin films to be layered on any surface through simple solutions (spin-coating, spray coating, inkjet printing, slot-die coating, and blade coating). These deposition methods can be easily incorporated into the roll-to-roll fabrication of flexible PSCs.

Consequently, it is vital to attain excellent control through the reaction between the inorganic and organic components, which leads to formation of perovskites with useful properties and device performance. Several process parameters such as deposition techniques, stoichiometry, thermal treatment, solvent engineering and additives which were incorporated into the process to enhance the perovskite thin film qualities will be discussed in the next section.

3. Pathway to perovskite thin film formation

The preparation of the perovskite layer before deposition on a substrate is fundamental for achieving high-efficiency devices. Thin film quality indicators, including crystallinity, grain size, presence of impurity phases, and defects, can influence the optoelectronic properties immensely. Furthermore, all the needed properties (like simple preparation, absorbance through the entire visible range) PV material are deemed to be affected by the quality of the film. The following parameters influence the quality of the perovskite film:

- Precursor stoichiometry
- Solvent engineering
- Annealing process
- Deposition method

Usually, careful control of these parameters during preparation of perovskite material is required to manufacture high-efficiency PSCs. The charge separation efficiency, recombination dynamics, and long diffusion length of the charge carriers are expected to be influenced by crystallization behavior and morphological evolution of the perovskite film for example, the large grain size can promote better charge transport.

3.1 Precursor stoichiometry

During the last decade, notable development has advanced the field of perovskite PV cells to the vanguard of thin film PV technologies. The stoichiometry of the perovskite material is of great significance as it predetermines the optoelectronic properties of the absorber and thus the device performance.

The stoichiometry of the precursor solution, specially the ratio of the organic and inorganic constituents, highly influences the resulting thin film quality in term of film conformity and carrier behavior. Moreover, precursor solution stoichiometry can effect profound changes in the properties of the perovskite layer along with the performance and stability of perovskite PV devices.

The initial nominal stoichiometry and non-stoichiometry of methylammonium iodide (CH_3NH_3I, MAI) and lead halides can be carefully regulated in both solution and vacuum evaporation techniques. The molar ratios of 1:1and 1:3 have been the most commonly used for perovskite precursors. But other molar ratios which are non-stoichiometric are reported in [7, 8].

Generally, a solution of PbI_2 and MAI with a stoichiometry of 1:1 is employed as the precursor to form a pure perovskite phase. For the mixture of $PbCl_2$ and MAI solution, a non-stoichiometric mole ratio of 1: 3 is used for the formation of $MAPbI_3$.

Any deviation from stoichiometric ratio ($PbCl_2$: MAI) can lead to either excessive MAI or unreacted $PbCl_2$ to remain inside the film, which can be expected to decompose their resulting optoelectronic performance. For example, the MAI and PbI_2 (1: 1 mole ratio) and the mixture of $PbCl_2$ and MAI solution, a non-stoichiometric mole ratio of 5: 3 is demonstrated to be optimum for the formation of perovskite precursor [8]. It should be noted that the mixed halide shows enhanced film properties, for instance. More compact and improved grain size, in comparison with the pure iodide $MAPbI_3$ perovskite.

3.2 Solvent engineering

The selection of the precursor solvent plays a key role in any solution method and controls the quality of hybrid perovskite films. In one-step deposition technique, a blend of two separate precursors (for instance methylammonium iodide and lead halides) is employed. Due to the distinctive nature of organic and inorganic precursors, the fundamental specification is that the solvent must be polar so as to dissolve the precursors, while the physical properties, such as boiling point and vapor pressure, must to be determined with respect to the preferred crystallization mechanism, that is, rapid or slow. Currently, γ-butyrolactone (GBL), dimethyl sulfoxide (DMSO), N-Methyl-2-pyrrolidone (NMP), and dimethylformamide (DMF) are the usual solvents in the precursor solution, and their mixture is a possibility to enhance the crystallization [9, 10]. It should be noted that addition of DMSO to the DMF solution aided in monitoring the perovskite crystallization, resulting in improved PV performance.

3.3 Anti-solvent treatment

In addition to the significance of the perovskite precursor–solvent interaction so as to decipher the nature of an intermediate phase in the solvent engineering technique, the application of an anti-solvent likewise plays an important role in the surface morphology and crystalline properties of the perovskite films The nature of an antisolvent is non-polar solvent (that is, a solvent in which the perovskite is not soluble). Consequently, the inclusion of the anti-solvent onto the precursor deposited on a substrate can generate fast supersaturation and precipitation of the perovskite precursor film by eliminating solvents from the precursor solution, hence, facilitating uniform grain growth with high nucleation density.

Olaleru et al. recorded that dropping ethyl acetate on a perovskite precursor film during the spinning process caused a homogenous perovskite film with good reproducibility [11]. Cohen et al. investigated the relationship between anti-solvent (toluene) treatment and electronic properties of the perovskite films together with surface morphology [9]. The most common antisolvents include toluene, chlorobenzene, and diethyl ether. Other antisolvents investigated including toluene, chlorobenzene, methanol, ethyl acetate and ethanol were found to induce fast crystallization and result in enhanced film quality [9]. Furthermore, the application of a more eco- friendly anti-solvent replacement to toxic chlorobenzene and toluene is greatly needed to minimize health risks.

3.4 Additives

The introduction of little amounts of chemical additives in perovskite precursor solutions have been recorded to offer advantages with regard to crystallinity, film uniformity and the resulting device performance. Additives constitute one more parameter affecting the perovskite morphology. There are different additives reported in the literature for various perovskites. Additive engineering can also enhance the quality of perovskite by modifying the film formation kinetics from a diverse perspective, including morphology modification, defect reduction, and enhancement of cation diffusion into PbI_2 layer [9].

It has been largely documented that the application of additives to the perovskite precursor solution could be a beneficial and logical technique to obtain large size grains in perovskite crystal, defect suppression, and improved stability and ultimately enhanced PCE of PSC devices. A wide spectrum of chemical additive has been recorded depending on the perovskite mixture and the additives manufacturing technique.

3.5 Annealing temperature

The formation of perovskite structure requires sufficient heat treatment whose purpose is twofold; removal of excess solvent remaining on the film and speeding up crystallization of perovskite structure. Annealing process is one of the major parameters controlling the perovskite formation and morphology. The perovskite annealing is performed at various temperatures; the maximum thermal annealing for $CH_3NH_3PbI_3$ perovskite should not be more than 110°C.

From our work [8, 11], the annealing temperature had strong impact on the formation of perovskite. The under-and over-annealing were both detrimental to cell performance. Under-annealing could cause incomplete conversion to the perovskite and leave residual solvent and ions in the layer, while over-annealing decomposed the perovskite layer into PbI_2 by evaporating MAI from the layer. Therefore, it was necessary to keep annealing temperature below 110°C to avoid PbI_2 formation and annealing time should be kept below 30 min.

A polar solvent including γ-butyrolactone (GBL), dimethyl sulfoxide (DMSO) dimethylformamide (DMF) and N, N-dimethylacetamide (DMA) in perovskite ink deposited on a substrate was removed by annealing temperature on a hot plate or drying in air after the spinning procedure, resulting in supersaturation in the film for nucleation and crystal growth of the perovskites. This technique provides an easy strategy to evaporate solvent for perovskite deposition.

4. Scalable fabrication of perovskite thin film

From a technological viewpoint various optoelectronic devices depend on the integration of active thin films in appliances such as solar cells, light-emitting devices, and transistors. Until now, vapor and solution deposition have been the two common routes/methods for effectively preparing hybrid perovskite thin films. One of the merits of hybrid perovskite over in-organic semiconductors is the feasibility of obtaining high quality semiconducting films by simple solution or sublimation processes.

The fabrication of a smooth and pinhole-free large-area perovskite thin film is essential in the quest to attain PSCs' commercial application. The main challenge relies on how to prepare top quality large-area perovskite thin films with a facile scalable deposition method.

Up till now, the understanding of crystal growth processes and precursor solution has resulted in notable development laid down on various scalable deposition methods. This section outlined the striking progress in the preparation of large-area perovskite thin films and PSCs with scalable deposition techniques.

The best route for the deposition of perovskite thin film in the laboratory is spin coating with the aid of antisolvent and additive engineering; but, this technique throws away perovskite solution and is very susceptible to numerous technical requirements, which stifles its selection for the fabrication of large scale perovskite films.

Until the present, the recorded techniques for the fabrication of large-area perovskite thin films ($1cm^2$) could be mostly divided into the following classes: spin coating, casting, blade coating, spray coating and slot-die coating technologies [12]. Although spin coating has intrinsic weakness for large-area film preparation, it is further mentioned in comparison to other methods [12, 13], which reflect the deficiency in preparing large-area perovskite thin film.

4.1 Spin coating

Spin coating is an easy and common technique widely employed for perovskite thin film deposition in the laboratory. Normally, a little quantity of precursor solution is dropped on the center of the substrate. Then, the substrate is spin at a high speed so as to spread the coating material by centrifugal force. The precursor solvent is normally volatile and concurrently evaporates during high-speed spinning. Consequently, the dried film with certain thickness is obtained.

Usually, there are two common methods for spin coating deposition of perovskite films. These are; one-step or two-step processing techniques as demonstrated in **Figure 2**. Perovskite thin film can be prepared either by spin-coating a mixed powders of MAX (X: I, Br, Cl) and PbX_2 (X: I, Br, Cl) solution (one-step coating) or by spin-coating PbX_2 accompanied by deposition of MAX (two-step coating).

Majority of current study on perovskite solar cells center on their fabrication using spin coating, which is already proven to attain better thin films with pinhole-free uniformity and carefully controlled crystal growth. Nevertheless, the spin coating is applicable to the manufacturing of non-scalable solution-processed PSCs [12]. It is difficult to expand the technique to enable fabrication of thin film for application in larger area devices. This practical bottleneck situation has strongly prompted scientists and manufacturers to consider new scalable techniques for the manufacturing of large-area perovskite thin films.

4.2 Vapor deposition method

The vapor-deposition techniques as displayed in **Figure 3** are usually performed under high-vacuum conditions. PbX_2 and MAX are deposited concurrently or alternately on the precoated substrate by thermal evaporation from dual sources of PbX_2 and MAX. Perovskite material is made at an adequate temperature and then crystallized into a perovskite film.

The perovskite film obtained from this strategy shows complete coverage, smooth grain structure with grain sizes up to micrometers and complete precursor transformation [14]. Indeed, it has been reported that vapor-deposited perovskite films were highly uniform with crystalline platelets at the nanometer scale [15]. This thermal evaporation method requires a high vacuum, which constraints cost efficiency and large scale production.

4.3 Casting

Casting requires easy transfer of perovskite ink to a target substrate accompanied by a brief heating procedure to evaporate solvent. Film thickness is regulated/influenced by the solid content in the solution and viscosity. The surface tension

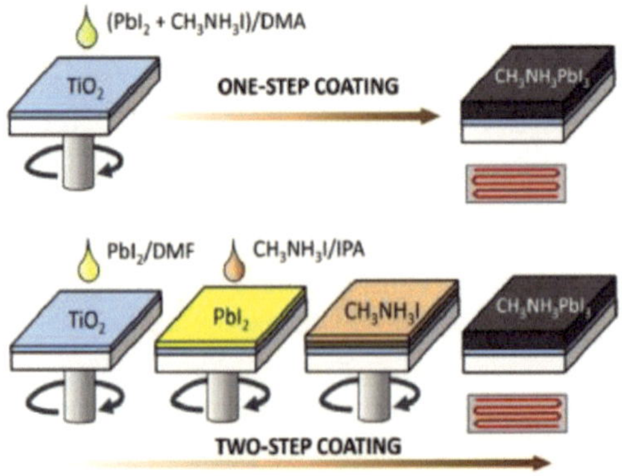

Figure 2.
Spinning coating process of perovskite thin film [12].

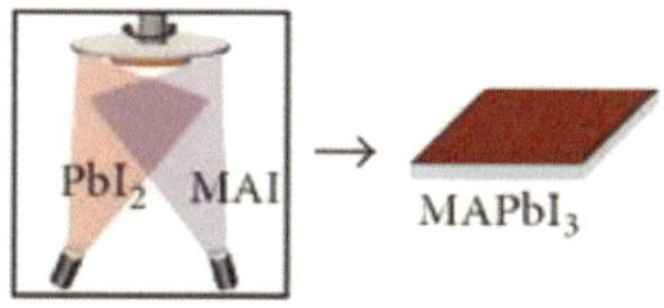

Figure 3.
Vapor deposition of perovskite layers.

between the perovskite ink and substrate together with perovskite precursor viscosity and annealing temperature will greatly influence the quality and uniformity of the film. The notable benefit of this method is that there is no need for specific equipment during the process. In this procedure, it is feasible to fabricate large-scale perovskite thin film, however, the inadequate monitoring of film thickness together with the imprecise perovskite ink employment are apparent weaknesses of this technique.

Currently, there is scant literature on fabrication of large-scale perovskite thin film with the casting technique [12]. The schematic illustration of this technique is demonstrated in **Figure 4**.

4.4 Blade coating

Blade coating is a highly established, scalable and extensively employed method preferred in fabrication of perovskite thin film for PSCs. It is a promising deposition technique due to little material waste, facile operation, and compatibility with roll-to-roll fabrication [16]. Blade coating was initially employed for manufacturing of dye-sensitized solar cells [16]. But recently, it has been applied to the deposition of large-area perovskite films after skillfully amended [12].

During the blade coating process, a sharp blade is positioned in an appropriate place over the target substrate with the precursor solution in front of the blade as displayed in **Figure 5**. The blade can be moved with linear speed within the substrate to obtain a wet perovskite film. Generally, the thickness of wet layer should be approximately half the distance between the substrate and the blade. The thickness and the quality of the blade coated film is usually ascertained by different physical factors such as surface energy of the substrate, surface tension and solution viscosity [16].

It has been reported that the blade coating generates a uniform wet film from the precursor solution through one-step deposition [18]. Deng et al. made large-area (1 $inch^2$) perovskite film with the doctor blading procedure and it was discovered that the carrier diffusion length of perovskite film obtained was longer than that of the films prepared by spin coating [19].

Figure 4.
Pictorial representation of hot casting and the inset is hot casting procedure [12].

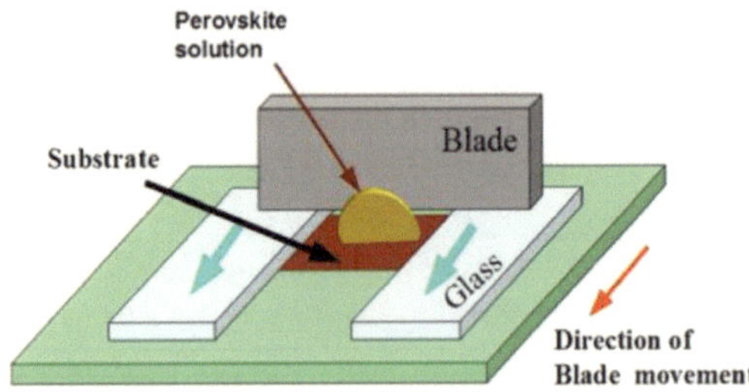

Figure 5.
Schematic of the blade coating technique [17].

4.5 Slot-die coating

Slot-die coating is another widely employed scalable technique for the fabrication of large-area perovskite thin film. It is probably the best fit technology for fabrication of perovskite thin film in manufacturing. Furthermore, the procedure is quite easy and relevant to a wide range of inks with various viscosities.

In this procedure, the precursor is usually applied from the coating head in the direction of substrate as shown in **Figure 6**. The perovskite film could be obtained, after drying and annealing. Quick and precise coating applied to the slot die is beneficial for the preparation of large-scale with high performance [16].

This method is far more expensive for laboratory uses in comparison with the blade coating. For this reason, the slot-die technique is not used often in research labs and the slot-die coated devices have much lower PCE than blade coated PSCs. Outstanding performance has been recorded so far on slot-die-coated PSCs. That is why, slot-die coating technique has been recommended as a roll-to-roll compatible process for the fabrication of PSCs in recent years by different researchers [12]. The basic illustration of the slot die coating technique as shown in **Figure 6**.

4.6 Spray coating

In brief, the ultrafine droplets of perovskite solution are sprayed from the nozzle and deposited on the target substrate which may be heated on a hotplate at certain temperature as demonstrated in **Figure 7**. After drying and annealing process, the perovskite film may be formed on the substrate.

Spray coating is a valuable low-temperature thin film deposition technique that allows a uniform, high coating rate applicable for various substrates. This method likewise draws great interest in the preparation of perovskite thin films [16]. But, little research on perovskite has been recorded employing the spray coating technique, maybe due to the instability of PSCs.

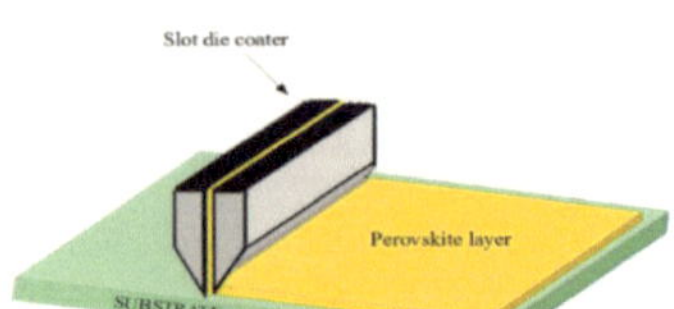

Figure 6.
The schematic illustration of the slot die technique for perovskite layer deposition [17].

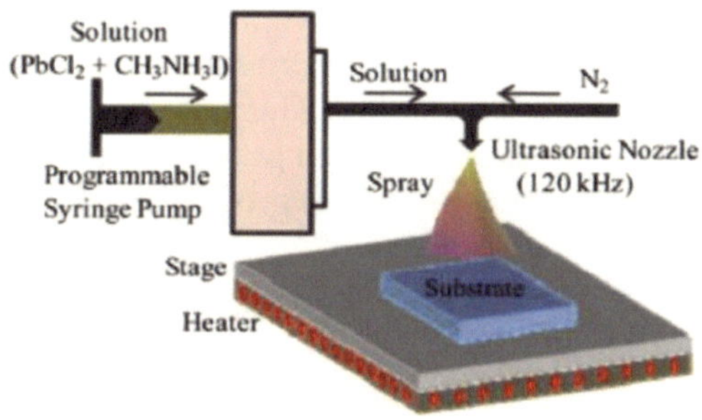

Figure 7.
Schematic diagram of ultrasonic spray-coating procedure [12].

In comparison with casting, spray coating has a better thickness control. In order to achieve a perovskite thin film with better quality by spray coating, the following should be considered; the perovskite solution concentration and viscosity, substrate temperature, the nozzle diameter and spray flow rate.

4.7 Ink-jet printing

An inkjet printing technique allows quick and material-conserving coating of thin films for the large-area preparation of photovoltaic thin films with the capacity to generate desired thin film as shown in **Figure 8**. However, to enhance the printing precision of the film the careful control of crystallization of the deposited wet films and solution properties is a fundamental condition. In this printing method, careful monitoring of the drop size is vital to obtain uniform thickness film. The ink properties take a leading role in assuring the stable droplets formation at the nozzle. Moreover, the fabricated film quality largely depends on the scattering of drops, coalescence, crystallization, and uniformity of the film during annealing. The parameters stated above are primarily influenced by the interaction of ink droplet with the substrate, for instance, the contact angle and surface roughness [20]. The astonishing success greatly promotes the application of this preparation method in perovskite devices.

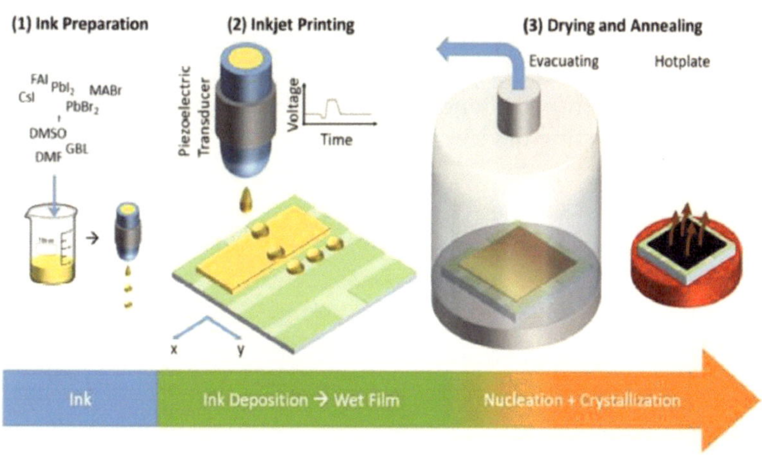

Figure 8.
Schematic illustration of the fabrication of perovskite material using ink-jet printing process [16].

Though, all these techniques are varying significantly from one to the other, but they have common goal of strengthen perovskite film formation so as to improve film quality and better device performance. The understanding of these mechanisms for creating new strategies is the key in producing top quality perovskite thin film. Further investigation of perovskite growth mechanism will be helpful in identifying opportunities for enhancement.

In consequence, optimizing processes will give more understanding into film formation, creating learning loops for developments in science.

5. Designing perovskite solar cells and modules

Perovskite thin films have yielded outstanding photovoltaic performance and are regarded compatible with low-cost, scalable fabrication. In spite of huge development in small-area PSCs' photovoltaic performance, the large-area fabrication of uniform and top-quality perovskite films remained problematic coupled with cell instability and toxicity. But, researchers are making progress meaning that there is hope that all the challenges will be resolved in the foreseeable future.

In order to establish the possibility in the direction of commercialization, it is a definite measure for emerging PV technologies to show high efficiency at a module/mini module of minimum 10–100 cm^2 [16].

Table 1 shows the record of various scalable deposition strategies employed to develop large-area perovskite thin films and their roll-to-roll capabilities [16].

The simplicity of making PV cells through solution-based and vapor-based processes as well as deposition techniques such as spin coating, spraying, slot die, blade coating and ink-jet printing make it possible for PV cell manufacturers to ultimately replace clean rooms and complex industrial equipment with facile bench top procedures. The rapid development, which has propensity for more improvements, coupled with low-cost materials and simple preparation processes. This will prompt a novel design, new reports on working mechanism and applied perovskite materials that will emerge.

It is therefore relevant to explore the best mix of various deposition techniques forming hybrid techniques. Though, blended techniques may likely give rise to dissimilar film formation mechanisms as a result of different interfaces such as solid/vapor or solid/liquid interfaces in relation with liquid/liquid interface in a sequential solution method. This calls for thorough investigations in the nearest future. The aim is to merge the benefits of various coating techniques for superior film formation and utilizing the benefits for the fabrication of large scale perovskite thin film for light-harvesting devices.

Method	Material	Roll to roll	Largest coaling area (cm^2)	PCE (%) single cell area ≤ 1 (cm^2)	PCE (%) module area ≥ 10 (cm^2)
Blade coating	$MAPbI_3$	yes	6 × 15	20.3	14.6
Slot die coating	$MAPbI_3$	yes	12.5 × 13.5	~12	10
Spray coating	$MAPbI_3$	yes	7.5 × 7.5	16.9	14.2
Ink jet printing	$MAPbI_3$	yes	2.02	17.74	—

Table 1.
Record of various scalable deposition strategies employed to develop large-area perovskite thin films.

6. Conclusion

Due to their 'adaptable' low band gap, the perovskite thin films demonstrated better advantages over their counterparts by far more absorbing over a broader range, facilitating full light absorption in films as thin as 300–500 nm, which is favorable for the solid-state cells. Perovskite absorbers find application in tandem with other PV technologies: supplementing spectral absorption, bandgap tunability, and simple technique and process compatibility with silicon and CIGS technologies. Huge progress has been demonstrated in efficiency, stability, and scalability. To become a market-competitive technology, a few number of issues such as toxicity must be resolved alongside with the feedback between process development, scaling, efficiency and stability.

In sum, the performance of PSCs can be enhanced greatly via the optimization of the different materials composition and techniques entailed in their development process. Particularly, for the fabrication of scalable modules using commercial solution coating equipment, the incorporation of solvent engineering and chemical additives together with physical qualities in the precursor solution is very much needed. Additionally, perovskite is not just for PV panels. These perovskite materials are excellent nanomaterials with fascinating properties and provide themselves to numerous inventions for novel devices. In consequence, perovskite's potential finds applications in the following: sensors, catalyst electrodes, fuel cells, lasers, sensors, memory devices and spintronic.

Author details

Akin Olaleru[1*], Eric Maluta[2], Joseph Kirui[2] and Olasoji Adekoya[1]

1 Physics Department, Yaba College of Technology, Lagos, Nigeria

2 Physics Department, University of Venda, Thohoyandou, South Africa

*Address all correspondence to: akinolaleru@gmail.com

References

[1] Joel Jean et al., 2015. Pathways for solar photovoltaics. Energy Environ. Sci., 8, 1200. https://doi.org/10.1039/C4EE04073B

[2] Jiandong Fan, Baohua Jia, and Min Gu 2014. Perovskite-based low-cost and high-efficiency hybrid halide solar cells. Photo. Res. 2 (5). https://doi.org/10.1364/PRJ.2.000111.

[3] N Qi Chen et al., 2015.Under the spotlight: The organic–inorganic hybrid halide perovskite for optoelectronic applications. Nano Today .10, 355—396. https://doi.org/10.1016/j.nantod.2015.04.009.

[4] https://www.frontiersin.org/articles/10.3389/fmats.2021.717830/full

[5] Shi, D., Adinolfi, V., Comin, R., et al., 2015. Solar cells. Low trap-state density and long carrier diffusion in organolead trihalide perovskite single crystals. Science 347, 519. doi: 10.1126/science.aaa2725.

[6] Nie, W., Tsai, H., Asadpour, R., et al., 2015. High-efficiency solution-processed perovskite solar cells with millimeter-scale grains. Science 347, 522. DOI: 10.1126/science.aaa0472.

[7] Longo G, et al. Journal of Luminescence189(2017)120-125

[8] Olaleru S.A, Wamwangi, D. et al. 2020. The impact of synthesis techniques on the properties of hybrid perovskite materials for photovoltaic application. Mater. Express, Vol. 10, No. 7, doi:10.1166/mex.2020.1747.

[9] Bat-El COHEN et al. 2016. Parameters that control and influence the organo-metal halide perovskite crystallization and morphology. Optoelectron., 9(1): 44-52. DOI: 10.1007/s12200-016-0630-3.

[10] Minsu Jung et al. 2019. Perovskite precursor solution chemistry: from fundamentals to photovoltaic applications. Royal Society of Chemistry. https://doi.org/10.1039/C8CS00656C.

[11] Olaleru, S.A; Kirui, J.K et al, 2021. Enhanced optical properties of perovskite thin film through material optimization for photovoltaic application. E3S Web of Conferences 239, 00020. https://doi.org/10.1051/e3sconf/202123900020.

[12] Zhichun Yang, Shasha Zhang, Lingbo Li, Wei Chen, 2017. Research progress on large-area perovskite thin films and solar modules. J Materiomics 3 231e244. https://doi.org/10.1016/j.jmat.2017.09.002

[13] Afzaal, M., Salhi, B., Al-Ahmed, A., Yates, H.M ., Hakeem, A.S., 2017. Surface-related properties of perovskite $CH_3NH_3PbI_3$ thin films by aerosol-assisted chemical vapour deposition. J Mater Chem C .5:4946e50. https://doi.org/10.1039/C7TC02968C.

[14] Peng Gao, Michael Grätzel , Mohammad K. Nazeeruddin . 2014. Organohalide lead perovskites for photovoltaic applications Energy Environ. Sci. https://doi.org/10.1039/C4EE00942H.

[15] Jin Cui, Huailiang Yuan, et al. 2015. Recent progress in efficient hybrid lead halide perovskite solar cells. Sci. Technol. Adv. Mater. 16 036004 (14pp). DOI: 10.1088/1468-6996/16/3/036004.

[16] Mazhar Abbas, Linxiang, Zeng et al., 2020. A Critical Review on Crystal Growth Techniques for Scalable Deposition of Photovoltaic Perovskite Thin Films. Materials 2020, 13, 4851; doi:10.3390/ma13214851

[17] H. Tyagi et al. (eds.), *Applications of Solar Energy*, Energy, Environment, and Sustainability, https://doi.org/10.1007/978-981-10-7206-2_16

[18] Qiu, S.; Xu, X.; Zeng, L.; Wang, Z.; Chen, Y.; Zhang, C.; Li, C.; Hu, J.; Shi, T.; Mai, Y.; et al. 2020. Biopolymer passivation for high-performance perovskite solar cells by blade coating. J. Energy Chem. , 54, 45-52.

[19] Deng YH, Peng E, Shao YC, Xiao ZG, Dong QF, Huang JS. 2015. Scalable fabrication of efficient organolead trihalide perovskite solar cells with doctor-bladed active ayers. Energ Environ Sci. 8:1544e50. https://doi.org/10.1039/C4EE03907F.

[20] Pulli, E.; Rozzi, E.; Bella, F. Transparent photovoltaic technologies: Current trends towards upscaling. Energy Convers. Manag. **2020**, 219, 112982. DOI:10.1016/j.enconman.2020.112982